Sheep Ailments

Two questions I am often asked.
A. How long can a ewe survive like this?
The answer is: 24–36hr on average.
B. What actually causes death?
The answer – heart failure due to a hypostatic
pneumonia which causes the lungs to fill with fluid.

Sheep Ailments
Recognition and Treatment

Eddie Straiton
The original TV vet

The Crowood Press

First published in 1972 by
Farming Press Limited as
*The TV Vet Book for Sheep
Farmers*

Sixth edition 1992, retitled as
Sheep Ailments

This edition published in 2001 by
The Crowood Press Ltd
Ramsbury, Marlborough
Wiltshire SN8 2HR

British Library Cataloguing-in-Publication Data
A catalogue record for this book is available from the British Library.

ISBN 1 86126 397 X

Acknowledgement is made to Mr W. A. Watson for the photograph of
bright blindness on page 149, and to Mr Ian Shaw, Veterinary
Investigation Officer, Worcester for the photograph on page 140,
on Border disease, Crown copyright.

Typeset by Textype, Cambridge
Printed and bound in Singapore by Craft Print International Ltd

Contents

Preface

One bottle of lambing oils, one bottle of fever drink, a piece of rolled-up fencing wire, pliers and a hammer. This was my equipment to fight disease and losses in a Scottish hillside flock during the early spring of 1938 – this and the limited experience of two previous lambing seasons. The shepherd was an alcoholic, and I had to cope almost entirely on my own. The wire, pliers and hammer were my 'stitching tackle'.

Looking back with present-day knowledge, it seems miraculous that my losses were not catastrophic: in fact they were comparatively slight. The only possible explanation must have been the natural resistance of a completely self-contained flock, plus the permanent pasture land.

Today things are quite different. The modern flockmaster, with his local knowledge, can embark on a lambing season with a veritable armoury of preventative drugs, so comprehensive that it should be a disgrace to lose a single ewe or lamb.

It is my hope that this work will serve as a simple, easily understood reference book not only for such flockmasters, but also for generations of agricultural and veterinary students and for veterinary surgeons in practice.

I acknowledge with sincere thanks the work and publications of the veterinary scientists, without reference to which it would not have been possible to present such a comprehensive disease picture. And once again, I express my appreciation and thanks to my photographer, Tony Boydon, and all the farmers who co-operated in providing so many of the illustrations.

EDDIE STRAITON
Stone, Staffordshire

Acknowledgements

My thanks also go to: Dr G. C. Coles, M.A., PhD, CIBiol., FIBiol. who very kindly edited the section on parasites; David Henderson of the Moredun Foundation for his co-operation in providing colour illustrations; the Central Veterinary Laboratory for the illustrations of a dropsical lamb and lamb dysentery; the Moredun Foundation – the ewe's eye is filmed over from contagious ophthalmia; the Scottish Agricultural College – strawberry foot rot; the Royal (Dick) School of Veterinary Studies – redfoot – also through the courtesy of Dr Phil Scott – the slides of uterine and vaginal prolapse; the Veterinary Investigation Centre, Shrewsbury – through the courtesy of Dr C. J. Lewis – the slide of orf affecting a ewe's nostrils; and the British Veterinary Association for permission to use their copyright of the orf illustration.

Foreword

By Dr J. A. WATT, MRCVS, PhD, BSc, Head of Veterinary Department and Director of Veterinary Research Edinburgh and East of Scotland School of Agriculture

As one who has long been involved in diagnosing disease and advising farmers and shepherds on health and flock management, it gives me considerable pleasure to write a Foreword to this welcome addition to the subject of health in the sheep flock. The approach is one which has been followed by the author in his previous work, that is, the liberal use of photographs to illustrate recognizable conditions and operations, accompanied by a clear, condensed text which demonstrates the author's wide practical experience.

The format of illustration and readable text should appeal in particular to stockowners and students, while avoiding the pitfall of producing 'do it yourself' experts. The importance of the sheep in agriculture tends to be underestimated, nevertheless in some areas in Britain it is of major and increasing importance in the livestock industry.

There can be few species in which so much attention has been directed towards the conception of flock protection and health, as opposed to conditions in the individual animal, and this wealth of epidemiological information has of itself generated the problem of dissemination and application of this knowledge. It has long been recognized that this education field, i.e. the application of results of research on the farm, is a difficult one, and for this reason material presented as a readable and practical reference work performs a valuable function.

This book, then, fills an important gap in the series so ably presented by the author, and its unique form of presentation will result in the dissemination of information in many quarters where the stereotyped textbook has little appeal.

J. A. WATT
South Queensferry, West Lothian

General Commonsense Advice

A veterinary investigation officer friend of mine, Mr Ian Shaw, gave me some of the soundest advice on sheep I've ever had. He said, 'When sheep are not thriving, wherever they are and whatever the conditions, before looking for disease, obscure or otherwise, always examine the teeth' (*photo 1*).

Over many years in practice this advice has saved me much heart searching and embarrassment. I would advise all flock-masters to keep it in mind.

Examination of the molar teeth, particularly in adult sheep, is not easy but it can be done by using a proper sheep gag and a powerful torch.

Again and again I have solved the problems of an unthrifty ewe by removing a loose or deformed molar. Such ewes do not always salivate or drool. They merely become progressively thinner. Obviously such dental problems are more likely to occur in older sheep, and can lead not only to malnutrition but also to death from pregnancy toxaemia.

The first sign may be the odd thin ewe in a flock of approximately the same age. Closer examination may or may not show salivation, but often the cheeks are swollen with wads of partially chewed food (*photo 2*) and the breath smells foul. The incisor or front teeth may or may not be irregular but usually the main trouble is in the cheek or molar teeth.

Treatment is very much a job for the veterinary surgeon. He will assess the success rate and economics of treatment.

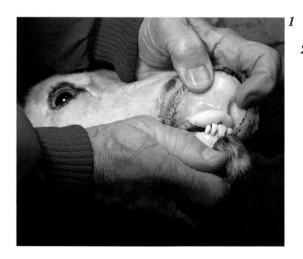

1

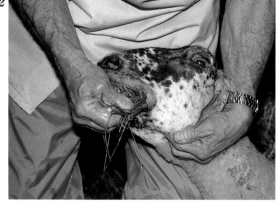

2

Cud-dropping (quidding) can occur where the teeth are normal.

In pedigree adult sheep the fault is sometimes thought to be hereditary but it is more likely to be caused by acidosis due to over-feeding. Quidding may also be associated with an infected wound of the tongue.

In lambs such a wound or infection should always be suspected; in fact a recent report suggests that the infection may occasionally be more or less identical to the 'wooden tongue' condition seen in cattle. The resultant cud-dropping can usually be cured by a course of antibiotics.

Explanation of sheep terms frequently used
A 'hogg' is a sheep that is no longer a lamb but has not yet been shorn; usually it becomes a 'hogg' around one year old.

After the first shearing the 'hogg' becomes a 'shearling'.

A 'gimmer' is a young female sheep. It can be a 'gimmer hogg' or a 'gimmer shearling'.

A 'teg' is the same as a hogg – a sheep between weaning and first shearing.

A 'wedder' or 'wether' is a castrated ram.

A 'teaser' is a vasectomized ram.

A 'hirsel' is an area of land which will support a flock of sheep – usually around 500 ewes – that can be looked after by one shepherd.

Metabolic Diseases

Perhaps the term 'disorders' is more appropriate since there are no germs involved in metabolic diseases.

They are produced by disturbances in the animal's metabolism, i.e. the conversion of food into the appropriate substances which are needed to keep the normal body functions going. The main organ concerned in metabolism is the liver which could be described as the 'factory of the body'. There the simple products of intestinal digestion are elaborated and built up before being transported to the various parts of the anatomy to provide heat and energy, and to build up and replace muscle, brain, etc. In order to perform this function the liver must be constantly supplied not only with the basic products of digestion – fatty and amino acids – but also with minerals, trace elements and vitamins.

Occasionally the 'factory' also has to play a part in breaking down stored food, chiefly body fat.

The metabolic diseases that concern us most in sheep are, first, pregnancy toxaemia, caused by excess ketones forming in the blood during the breakdown of body fat; secondly, hypomagnesaemia, due to a shortage of the mineral magnesium; and thirdly, lambing sickness which, like milk fever in the cow, is caused by a deficiency of calcium.

Strictly speaking, cerebrocortical necrosis (CCN) should also be included under this heading, as should swayback, pine (or cobalt deficiency), and phosphorus and vitamin E deficiencies.

1 Pregnancy Toxaemia (Twin Lamb Disease)

In the simplest possible language, pregnancy toxaemia is brought about by inadequate feeding or, to put it even more bluntly, by starvation. This is why the disease occurs when there is a deterioration in the quality and quantity of the diet during the last 2 months of pregnancy. Any additional stress can hasten its onset.

For example, it is seen in flocks maintained on pasture with little or no supplementary feed, and occasionally a few cases may occur on a good pasture during an open winter. It is also seen in severe weather where frosts and snows drastically reduce or completely cover up the available herbage (*photo 1*). It occurs worldwide.

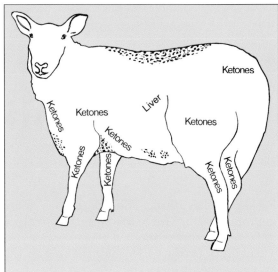

How it occurs

In late pregnancy, especially if twin lambs are present, the ewe obviously requires a good food intake to keep pace with the growth of the lambs. If she is not getting enough to eat, she has to turn to her own food reserves which are stored in the form of sugar or glycogen in the liver and muscles, and in the form of fat throughout the body.

The hungry ewe quickly uses up all the sugar reserves and this causes a drop in the blood sugar content. She then turns to the body fat; this is broken down in the liver, and during the breaking-down process, poisonous substances called ketones are formed (*see diagram*). These accumulate in the blood-stream and produce an effect not unlike that caused by excessive alcohol in humans.

Symptoms

The first sign is the ewe's disinclination to move, quite distinct from any disability. The animal is often described as being 'stupid', and this symptom of apparent stupidity, together with the history of pregnancy, is virtually diagnostic.

In this state the ewe will often face, and even try to fight off, the sheepdog (*photo 2*). At this stage a vet can confirm the diagnosis by testing a pin-prick of blood from the ewe's gum with glucose test strips.

Later the ewe's head is carried in an unnatural position, usually to one side (*photo 3*), and it may be held high or dropped low.

The gait becomes staggering and uncertain, and later still there is a disturbance of vision and even blindness. This is why in many parts of the country the condition is described as 'snow blindness' (*photo 4*).

The temperature is usually normal, but the

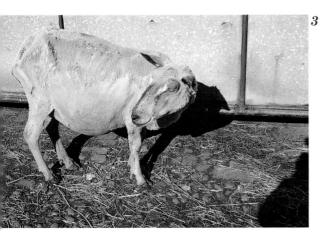

3

4

appetite is completely lost, and the patient is invariably constipated.

In about 1 to 2 days the ewe is unable to stand; she becomes progressively comatose and dies within 1 to 6 days.

Treatment

Treatment of cases is very unsatisfactory; unless the ewe is close to lambing the mortality rate can be up to 90 per cent.

My own rule of treatment is: if she is within a fortnight of lambing, I treat her by giving an anabolic steroid to stimulate the appetite and raise the blood sugar level and combine that with intravenous dextrose and vitamin B injections. At the same time I prescribe treacle and glycerol by the mouth.

If the affected ewe aborts she has a better

chance of recovery even if the lambs are born dead. It may pay to abort her, or to perform a Caesarian section.

If, however, lambing is some way off and no toxic treatment has been given (usually the best way to estimate how far the ewe is off lambing by examing the udder (*photo 5*), then I recommend loading up the ewe immediately and taking her to the butcher with the possibility of salvaging the carcase.

Quite obviously it is much better not to have to treat pregnancy toxaemia, and in the light of present-day knowledge, any intelligent shepherd can prevent it.

Prevention

The simple clue to the virtual elimination of pregnancy toxaemia in sheep lies in the feeding, particularly during the last 8 weeks of pregnancy when the major part of the foetal growth takes place.

Demand for extra food during this period depends on the number of lambs the ewe happens to be carrying. This can now be assessed by scanning; but failing that, ewes with a condition score of less than 3 should be separated for special attention. The alternative is to feed on the presumption that all the ewes are carrying twins.

The most important feed of all is good leafy hay, but this must always be provided ad lib (*photo 6, overleaf*). Again and again farmers moan that their ewes don't eat the hay they put out. This is nonsense. They don't eat non-

5

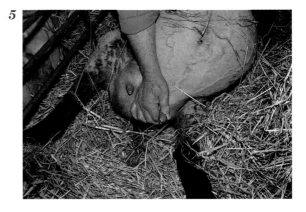

6

stop like cattle or horses, but they nibble away at regular intervals and over the 24 hours pack in an adequate supply, provided (and this leads me to my second, most important, point) ample fresh water is available at all times.

The pregnant ewe needs plenty of water, as does the lactating ewe and, for that matter, all other sheep. The stupid idea that sheep don't drink water is still being bandied about, often by apparently intelligent countrymen.

Good hay and plenty of water, ad lib, could be all that is required if the ewes are housed. In fact, in housed sheep, top quality hay will provide sufficient nutrients for the full growth of a single lamb and most average-sized twins. But if you rely on hay alone, make sure you have it tested for quality – only the best is good enough.

If the ewes are outside, then it is a different matter. Much of the hay ration is used up in keeping the ewe's body heat normal and a supplementary diet of concentrates becomes vital. The time to start the concentrates is 8 weeks before lambing, when the foetus or foeti really start to grow (*photo 7*).

I've found over the years that the quality of concentrates depends largely on the type of land and pasture. If the ewes have been on good pasture, then the concentrate ration should be at least as rich in protein, probably 1 or 2 per cent richer than the best grass on the farm.

If the land and pasture are comparatively poor, then a cereal concentrate of crushed oats and flaked maize will do admirably, but keep barley out of the ration. It has been my experience that barley can do as much, if not more, harm to pregnant ewes than it does to cattle.

As an additive, I always recommend diluted treacle. It makes the feed attractive and succulent and has the obvious benefit of being ketogenic, providing readily assimilable carbohydrate. Ketonic licks are also excellent (*photo 8*).

In all hypomagnesaemia areas, added magnesium is vital. I advise feeding at the rate of 7g (¼oz) of calcined magnesite a head each day (sheep find it most edible if wet beet pulp can be included in the ration), and I feed it right on to the middle of June.

As for other minerals, their necessity depends on the land and on the standard of farming. Certainly in my comparatively rich, well-farmed area the only trace element I

8

7

advise is a 5 per cent copper lick for farms with a history of swayback. There is no doubt that oral copper in any quantity is highly dangerous to all sheep, and the weak lick is the only safe way I've found of feeding it.

It is my opinion that considerable amounts of money are wasted in feeding minerals where they are not necessary. After all, good hay, which I stipulated at the start, contains all the minerals and trace elements in the correct quantities and in an easily assimilated form. Excess minerals are not only wasteful, but they can also damage the liver and intestinal tract of the ewe and lead to diarrhoea and death.

Lastly, the quantity. Having compiled the ration carefully in consultation with your veterinary surgeon or your local advisory officer, and having added magnesium where necessary and any other vital element which is known to be missing from the soil and pasture of your respective farm, start 8 weeks before lambing with 225g (½lb) of your concentrate mixture for each ewe every day (in addition to the ad lib hay and water) and gradually increase the quantities, i.e. **keep the ewes on an ascending plane of nutrition**.

At 4 weeks before lambing, feed 450–675g (1–1½lb) per head a day, and at full term feed

9

1,125–1,350g (2½–3lb) according to the breed and size of your ewe flock. In a particularly severe winter it will pay to add 5 or 10 per cent of animal protein (fish meal or meat-and-bone meal) to the ration during the last 3 or 4 weeks.

Do all this and pregnancy toxaemia will occur only if or when you allow the water supply to freeze. *So never forget in severe weather to break the ice several times daily* (*photo 9*).

The routine technique of scanning the pregnant ewes allows special feeding for the ewes carrying twins or triplets and is very important not only in preventing pregnancy toxaemia but in ensuring an adequate milk supply for the lambs.

2 Hypomagnesaemia

Hypomagnesaemia is common nearly everywhere, especially in lowland sheep. Nowadays it is probably the greatest potential danger to all flocks.

The syndrome may appear under several circumstances. It can occur after a journey from high ground to lowland pastures, due probably to a 'transit tetany', with the magnesium deficiency appearing as a direct result of the travelling.

Hypomagnesaemia can also break out in a flock during a sudden cold spell, no doubt because of the extra strain placed on the sheep's metabolic rate.

Most cases, however, arise in exactly the same way as in cattle – i.e. by a direct shortage of the mineral magnesium. Like cattle, the sheep's magnesium reserves are stored on the crystalline surfaces of the framework of the bones, especially the ribs and vertebrae.

In young sheep the bony framework is open; the magnesium reserves are readily available and may last for 40 to 50 days – much longer than they do in older animals, where the bone structure is more compact. In fact, in adult sheep the magnesium reserves may keep them going for no more than 4 or 5 days. If the flock is on a low plane of nutrition, the magnesium reserves are soon exhausted.

On a good diet, however, hypomagnesaemia can, and does, occur quite commonly. Then it is probably triggered off by the hay (or silage, if fed) having been made early from young, artificially flushed grasses.

In any case, even if symptoms do not appear during the winter, there is always a gradual natural decline in the magnesium reserves, and consequently all the sheep are lowest in magnesium in the early spring.

If they are then turned out on to a young, rapidly growing pasture which owes its early growth to artificial fertilizer, the sheep can start to die by the score.

It appears that the fertilizer's nutrients, especially potash, lock the magnesium in the soil and hinder its uptake, or that the high nitrogen content of the artificially boosted grasses inhibits absorption in the ewe's intestinal tract.

Obviously, therefore, ewes are at greatest risk during the lambing period.

Symptoms

In my experience – usually sudden death. Spotted in the early stages the sheep may stagger about; the animal is hyperexcitable with twitching muscles and grinding teeth and, if untreated, soon falls down in a fit – kicking and frothing at the mouth (*photo 1*) before becoming comatose and dying.

Treatment

If caught reasonably early – even while the animal is in a fit – treatment can be

spectacularly successful. Often I have pumped 100cc of magnesium solution into a prostrate kicking ewe and within an hour have been unable to spot any difference between her and the rest of the flock. But, for such results, treatment must be given early because after a comparatively short time, haemorrhages occur in the brain and heart. When that happens, magnesium injections are useless.

Personally I like to use a 20cc syringe and inject the ewe under the skin in five different places (*photo 2*), coating the hypodermic needle with antibiotic between each one and massaging the sites to disperse the solution. In this way a much more rapid absorption and effect are obtained.

If the ewe is dead or dies, always get your veterinary surgeon to perform a post-mortem examination. The scientists say that hypo-

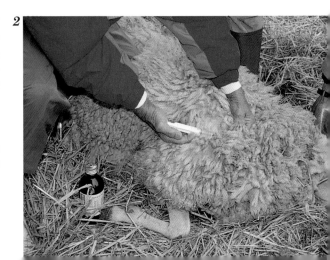

magnesaemia post-mortems reveal little, but I have found over the years that one common feature, which I have come to regard as diagnostic, has been well-marked haemorrhages in the substance of the heart muscle (*photo 3*).

Prevention

If the flock is at all suspect, it might be a good idea to have blood samples taken from, say, 10 per cent and checked for magnesium levels. However, I prefer to persevere with preventative therapy at the first sign of loss, and I think most practising veterinary surgeons do likewise. In fact I'm sure that the majority, like myself, will have the hypomagnesaemia farms earmarked and will advise prevention at all times.

Luckily most hypomagnesaemia cases are found during the winter and early spring. This is fortunate because at these times it can be controlled by feeding the magnesium. This can be done in several ways: (i) as a magnesium-rich mineral, which must contain at least 60 per cent magnesium to be effective, (ii) as sheep cubes or nuts containing magnesium, or (iii) as magnesium acetate in molasses, dispensed in freely available ball licks.

I prefer to feed 7g (¼oz) of calcined magnesite per head per day to the entire flock. The magnesite should be well mixed up with the concentrates, and ad lib hay and water should also be provided. Where possible, wet beet pulp or treacle added to the ration will make the calcined magnesite more palatable.

As a long-term policy in all the hypomagnesaemia areas, but especially on arable and certain hill farms, I advise the routine use of magnesium limestone instead of ordinary lime as a pasture dressing. This should contain at least 10 per cent of magnesium. Spread at 1 tonne per hectare (2½ ton per acre), magnesium limestone will increase the magnesium content of the pastures by as much as 25 per cent. The ideal dressings would be calcined magnesite – 1.25t/ha (10cwt to the acre) – or Keiserite – 110kg/ha (100lb to the acre) – but both are expensive.

One other commonsense prevention, especially in the spring, is to change the flock on to another pasture, preferably an old established turf which has had little, if any, artificial fertilizer applied (*photo 4*).

There is no doubt that when the disease flares up in a young ley, deaths will often cease dramatically if you do nothing else but simply shift the flock on to a permanent pasture. It seems that when magnesium is available, sheep can use it quickly.

If a client insists on maintaining an intensive grass programme, I advise alternate grazing of the young leys and of older permanent pasture, at least until the end of June. These rest periods allow the grasses in the leys to take up a percentage of the magnesium.

No doubt the eventual answer to hypomagnesaemia will be the discovery of the precise trigger factor which inhibits the magnesium uptake.

4

3

3 Lambing Sickness

Whereas hypomagnesaemia can be seen at any time, lambing sickness or hypocalcaemia usually occurs either just before, during, or after lambing (*photo 1*). The condition is identical to milk fever in the cow. Older ewes are more commonly affected.

Cause
It is caused by calcium deficiency. Considerable quantities of calcium are required to make the bones and teeth of the lambs. During and after the birth of the lambs the udder fills up with milk, which contains a fair percentage of calcium. Occasionally the ewe's thyroid gland cannot cope with this sudden additional demand and lambing sickness is the result.

Predisposing causes
Fatigue – for example, following the gathering in of hill ewes. Sudden changes in feeding can trigger it off, as can starvation.

Symptoms
The ewe may drop dead or fall flat out when being driven, but usually she goes off her food and her ears become ice-cold. The muscles start to tremor, the hind legs stiffen and she staggers about like a drunken person. She then goes down and is unable to rise. The breathing is accelerated and, if untreated, bloat sets in. The temperature is normal and constipation is the rule.

Treatment
A spectacular response is obtained with intravenous or subcutaneous injections of a 20 per cent solution of calcium borogluconate (*photo 2*). Again, 100cc is the dose and when given subcutaneously is best injected at five different sites using a 20cc syringe, coating

the hypodermic needle and the sites with antibiotic.

Prevention

Unfortunately there is no panacea for the prevention of lambing sickness. Where cases are prevalent it is wise to check the lime content of the pasture. Recently, after a severe run of trouble on my own farm, I found that one of my sheep pastures required 4 tons to the acre. On hill sheep farms, however, this may not be practicable.

There are certain other commonsense procedures, e.g. avoid sudden changes of feeding in late pregnancy. And do not drive the ewes too much immediately prior to, or immediately after lambing.

All shepherds, especially those on remote hill farms, should be provided with the necessary calcium solution, 20cc syringe and needles so that they can treat suspect cases immediately.

Needless to say, adequate calcium should be present in the diet of ewes in late pregnancy – though this should present no problems if concentrates are being fed and good quality hay is available ad lib, even in early spring (*photo 3*).

Many affected sheep also show a degree of hypomagnesaemia, so it is wise to use a calcium solution with added magnesium.

Ewes which recover from lambing sickness are more prone to losing their lambs than the unaffected members of the flock.

3

4 Cerebrocortical Necrosis

1

This disease, known as CCN, is a nervous disease of ewes and lambs, associated with necrosis (death) of part of the brain. It is seen mostly in lambs 2–4 months old, though younger lambs can also be affected (*photo 1*).

Causes

The precise cause is not known, but research work seems to indicate that it is due to a thiamine (vitamin B_1) deficiency caused by some metabolic disturbance. Certainly some cases, if spotted early enough, respond to treatment with injections of vitamin B_1.

Symptoms

In the early stages the affected ewes and lambs wander round aimlessly; they may walk in continuous circles or may just stand motionless. They also appear to be blind – staggering and swaying as the condition progresses.

After a few hours they pitch forward on their side or brisket, throw their heads back (*photo 2*) and kick their legs as though in a fit. The legs often stiffen, but any sudden noise will trigger off the violent leg kicking.

There are two other conditions which can produce almost identical symptoms, viz. lead poisoning and hypomagnesaemia, though I have found lead poisoning to be extremely uncommon in sheep. Nonetheless a specific diagnosis can only be made on post-mortem examination in a laboratory.

Treatment

It is unsatisfactory but all suspect cases should be injected with vitamin B$_1$ immediately, either intravenously or intramuscularly; at the same time it is wise to provide a cover of long-acting antibiotic to

prevent meningitis or other complications.

Since prompt differential diagnosis in the field is virtually impossible, it is wise to inject magnesium and also to give a solution of magnesium sulphate by the mouth – both will take care of any magnesium deficiency, and the oral magnesium sulphate will act as the antidote to possible lead poisoning.

Prevention

Once it is known that the condition has become established on a farm, a vitamin supplement containing thiamine should be fed to the entire flock.

5 Swayback

In Britain swayback is a disease of new-born lambs (*photo 1*), though in mild cases the typical symptoms may not be readily apparent for a number of days or even weeks. It is seen mainly where the soil type is of peat, limestone or clay.

Cause

It is caused by a copper deficiency in the ewe's diet, producing a low blood copper level. This low level often occurs where there is no copper deficiency in the pastures. It would appear, therefore, that there is a factor as yet

undiscovered which inhibits the uptake of the copper. Possible predisposing causes appear to be over-liming of the pastures or excess of sulphur and molybdenum.

Symptoms
When the blood copper of the pregnant ewe is low, the unborn lambs suffer a varying amount of damage to the brain cells.

The brain damage usually manifests itself at birth by the lamb having difficulty in standing or in controlling its hindquarters (*photo 2*). However, in milder cases, the symptoms may not appear until the lamb is a few weeks old; then the affected lamb may only sway almost imperceptibly (hence the name swayback) (*photo 3*). In fact, in very mild cases the hindquarter weakness may only be seen when the lamb is chased.

In certain parts of the world copper deficiency produces clinical disease in adult sheep characterized by loss of wool quality. Occasionally symptoms similar to those seen in British lambs, but without impairment of wool quality, are also observed.

Treatment
If a lamb is not too badly affected it is well worth injecting or dosing it with one of the modern 'safe' copper products, if only to avoid it getting worse. At least such compounds, properly used, will promote the lamb's growth rate.

Chance of recovery?
It has been my experience that where the symptoms are severe, the lamb never gets better. In fact, the swaying becomes progressively worse as the weight of the lamb increases (*photo 4*). However, I have seen some attacks mild enough to allow the lamb to be fed in the normal way. My advice, therefore, is: where the lamb can walk it is wise to give it a chance by trying the treatment.

Prevention
There is a choice of two preventative routines. First, where only an occasional case

occurs, provide 0.5 per cent copper licks to the ewe flock before and during pregnancy (*photo 5*). **The licks appear to be safe if they are**

5 **the only source of copper, and are apparently effective on some farms.**

Secondly, where swayback is or has been a major problem:

A. During an outbreak, dose all the pregnant ewes with one of the 'safe' copper products, taking care to follow the dosing instructions carefully. Both the injections and oral compounds are effective.

B. In subsequent years supply a controlled dose to all pregnant ewes in mid-pregnancy. **Other sources of copper such as minerals in the concentrates and copper licks should be removed.**

C. It may be necessary to reduce the dose in small sheep **but at all times be particularly careful because of the danger of copper poisoning** (see page 149).

D. Avoid the over-liming of the pastures.

6 Acidosis

This is an acute metabolic disease of sheep characterized by sudden death or more frequently by anorexia (i.e. off feed), marked depression and general weakness accompanied by tachycardia (i.e. a rapid heart beat) (*photo 1*). There may be signs of abdominal pain (i.e. the ewe may kick at her belly), and a diarrhoea covered or mixed with mucus may develop with the patient recumbent and ice cold.

Cause
Sudden change to, or the over-eating of, a grain feed, especially barley. In Britain it is seen in the autumn when lambs are brought indoors and suddenly put on cereals; or it can occur if store lambs are turned onto a barley stubble where, due to adverse weather conditions, the grain has not been properly

harvested. And of course it can occur at any time if sheep accidentally get into a grain store or if older hill sheep are brought down

to a lowland farm and fed cereals for the first time.

Treatment

In general practice I have found acidosis to be mainly a problem among cattle, but it is occurring more and more frequently in sheep. The treatment is the same as for cattle, viz. intravenous injections of concentrated vitamin B combined with drenching with a solution of sodium bicarbonate in cold water.

If bloat is present it can be relieved by giving a dose of nut (arachis) oil by the mouth – a wineglassful is usually sufficient.

The dehydration can be alleviated by giving one or two pints of glucose saline solution intravenously. Obviously these injections have to be given by a veterinary surgeon.

Prevention

Largely commonsense.

- Avoid barley feeding if at all possible.
- Introduce the cereals or concentrates gradually over a week or 10 days.
- Make sure that good quality hay is available at all times; also fresh clean drinking water.
- Provide plenty of trough space as an insurance against greedy feeders.

7 Pine

Pine is a chronic wasting disease of sheep. In some respects it is similar to pernicious anaemia in humans.

Cause

The basic cause is a shortage or more usually a complete lack of the trace element cobalt. The trigger factor, however, is the inability of the sheep to manufacture vitamin B_{12} in the rumen because of the complete absence of cobalt.

Symptoms

All the growing lambs lose their appetite and fail to thrive in summer despite regular dosing against internal parasites. The older sheep have dull, dry fleeces (*photo 1*) and become stunted, acutely anaemic and handle badly. If the condition is not diagnosed, emaciation continues until the sheep cannot stand, goes into a coma and dies.

Treatment

Fortunately the pine areas of Great Britain are now well known and prevention is more the rule. However, affected sheep respond rapidly to vitamin B_{12} injections plus dosing with a cobalt bullet (*photo 2, overleaf*). The cobalt bullet stays in the sheep's first stomach for about 6 months and keeps supplying the small amounts of cobalt necessary.

1

Prevention

Where pine is suspected, it is always wise to have the diagnosis confirmed by a research institute.

Prevention is not always easy, especially

2 on hill land. Cobalt can be spread on the pastures, fed as a trace element in a mineral mixture or given as a drench, but all these methods have obvious disadvantages.

Perhaps the best preventative method is the twice-yearly dosing of the individual sheep with cobalt bullets.

In Britain the pine areas are the coarse sandy soils of Scotland, the limestone areas of the Pennines, the old red sandstone areas of Hereford, Shropshire and Worcester and around Dartmoor, and the edge of the chalk in southeast England.

8 Vitamin E or Selenium Deficiency

This is another wasting disease of sheep which is also known as muscular dystrophy and white muscle disease (*photos 1 and 2*). It is seen mostly in young lambs during the first few months of life.

2

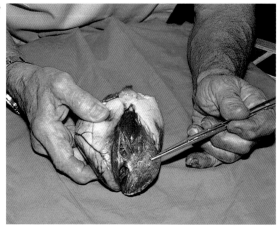

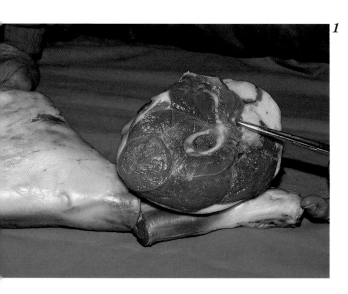

1

Cause

The specific cause is a shortage of vitamin E or selenium or both, but the general predisposing cause is usually starvation.

Symptoms

Acute and rapid wasting of the skeletal muscles despite a good appetite. The affected lamb is soon unable to stand. The heart muscle is also affected and death occurs within a few days from heart failure. A few animals may be found dead within hours or days of being turned out onto fresh spring pastures. (*Note.* The lambs may be born dead.)

Treatment

A veterinary surgeon will be required to make the diagnosis and he may well have to seek the help of a scientist. Post-mortem examination shows a characteristic wasting and whitening of the muscle fibres on the skeleton and in the heart.

Provided the condition is not too far advanced the affected sheep respond to vitamin E and selenium given by injection and by the mouth. More important, however,

is to prevent the rest of the flock going down with it.

Prevention

Improve the standard of nutrition immediately. Feed ad lib best quality hay and provide a concentrate mixture containing a vitamin supplement and selenium.

Any lambs at risk can be injected with vitamin E and selenium every three weeks till out of danger. Your veterinary surgeon, who should be consulted, will advise you on this.

Note. In certain parts of the world – chiefly North America and Australia – an excess of selenium can cause poisoning in one of two forms:

A. A chronic type called alkali disease.

B. An acute form with a high temperature, rapid breathing, prostration and death.

9 Phosphorus Deficiency

A mild phosphorous deficiency in the pregnant or lactating ewe can produce a condition not unlike lambing sickness or mild hypomagnesaemia. In fact any cases of lambing sickness, or suspected hypomagnesaemia, that fail to respond completely to straight calcium or magnesium very often recover spectacularly when given combined calcium, phosphorus and magnesium.

However, severe and persistent phosphorus shortage can lead to a chronic wasting disease, especially when protein and energy foods are also in short supply.

Cause

The usual cause in my experience is partial starvation on a phosphorus-deficient herbage: in other words, a deficiency of protein and energy as well as of phosphorus.

Symptoms

The condition occurs in young growing sheep from 4 months old onwards. The bones do not develop properly, and this gives rise to three main conditions: double scalp, rickets, and a disease known as open mouth.

Double scalp

This is seen towards the back-end, especially on some of the poor hill pastures of Scotland.

Symptoms

The affected sheep are poor, light and unthrifty. The bones at the front of the skull are thin and flexible (*photo 1, overleaf*): when you press these you can feel the inner layer of bone – hence the name 'double scalp'.

Treatment and prevention

It is not enough just to provide phosphorus: the overall plane of nutrition must be raised very considerably immediately. Change to a better pasture, and supply best quality ad lib hay and concentrates. The affected sheep will then hold their own and will begin to recover more or less completely during the subsequent summer grazing.

Incidentally, since all unthrifty sheep are particularly susceptible to gastro-intestinal parasites, it is wise to inject or dose with anthelmintics before changing the pasture.

Rickets (Bent-Leg)

This is due to a deficiency of phosphorus and vitamin D, combined with rapid growth.

Symptoms

It is seen in young, rapidly growing lambs (*photo 2*), especially when they are very well fed, e.g. young ram lambs being fed for sale or

when there is an excess of lush pasture in the spring. The rapidly growing lamb suddenly goes lame and one or more of the legs start to bend at the joints or in the shafts of the long bones (*photo 3*). The affected joints may be swollen and painful.

Treatment and prevention

Vitamin D injections, combined with calcium and phosphorus supplements during the time of rapid growth. A concentrate mixture containing 10 per cent of fish meal or 5 per cent of meat-and-bone meal will very often prevent further cases.

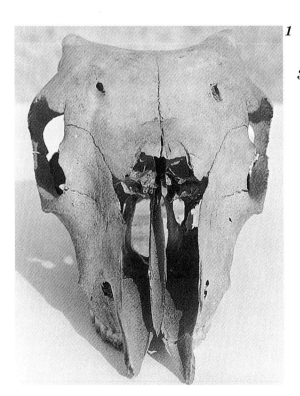

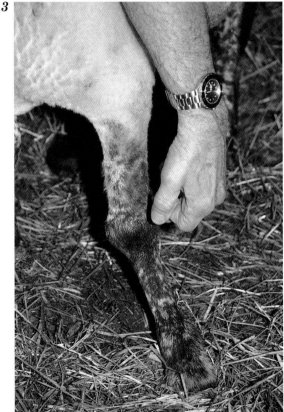

Open mouth

Again, a disease of young growing sheep. It appears to be almost a combination of rickets and double scalp and is due to a complex of both these diseases.

Symptoms

The affected lambs cannot close their mouths and so the incisor teeth do not reach the dental pad. Naturally they rapidly lose condition because they cannot graze properly.

The lower jaw bone is spongy and easily bent. Salivation is usually a feature.

Treatment

It is advisable to slaughter the affected animals.

Prevention

Concentrate on using the methods advised for double scalp and rickets. With all deficiency diseases the golden rule is to improve the plane of nutrition.

10 Vitamin A Deficiency

Deficiency of vitamin A can lead to a degeneration of the eye surface which produces a condition called xerophthalmia. It is easy to confuse such a condition with contagious ophthalmia (see page 141).

Symptoms

Large batches of the flock are attacked, simultaneously because the deficiency is common to all.

The eyes start discharging and quickly cloud over (*photo 1*).

If untreated, ulcers form and secondary infection can cause permanent loss of vision.

Treatment

Vitamin A injections plus a vitamin supplement in the concentrate feed; in addition to ad lib good quality hay.

The administration of antibiotic applications to the eye or subconjunctival injections of long-acting antibiotics will prevent secondary infection.

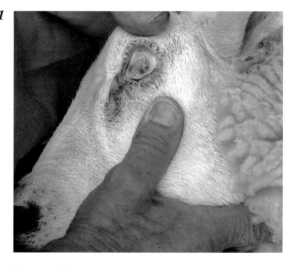

1

Prevention

Good feeding especially during the winter. Wherever possible a vitamin supplement should be added to the concentrates.

Without doubt the most common cause of all deficiency diseases is malnutrition or at the very least inadequate feeding.

Clostridial Diseases

11 Enterotoxaemia

When a sheep dies suddenly, the most likely cause of death is either hypomagnesaemia or enterotoxaemia.

Enterotoxaemia embraces three recognized conditions:
Enterotoxaemia of adult sheep (*photo 1*).
Pulpy kidney in lambs 3 to 12 weeks old.
Struck, seen in sheep over 1 year old.

Cause
A bacterium called *Clostridium perfringens*. There are two main types of this germ, Type D and Type C. Type D produces the enterotoxaemia of older sheep and the so-called pulpy kidney of lambs. Type C causes struck in sheep over the age of 1 year.

When and where enterotoxaemia occurs
Enterotoxaemia is an acute fatal disease occurring nearly everywhere and seen in sheep of all ages.

It produces pulpy kidney in lambs 3 to 12 weeks old, particularly single lambs.

It also hits weaned lambs and adult sheep, often affecting the biggest and best animals. In other words, it appears to hit the thriving lamb or sheep the hardest. The main danger period is the first few days after the flock is put on a changed or an improved diet, i.e. a rich pasture or a higher level of concentrates.

Clostridium perfringens Type C hits sheep over 1 year old which are grazing high quality pastures in the late winter and early spring, but mainly in certain areas of Britain, e.g. Romney Marsh, North Wales and the Lothians.

Given favourable conditions, Types D and C grow rapidly in the intestines (*photo 2*) and

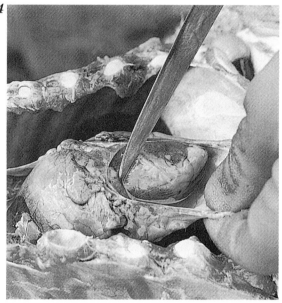

excrete a highly lethal toxin. This toxin is absorbed and causes the death.

Symptoms

Sudden death without symptoms is the rule, and I have seen many sheep and lambs die from this complaint. They fall down in a fit, throw their head back and kick furiously, apparently in abdominal pain (*photo 3*). They then lapse into a coma and die.

There is no treatment.

Post-mortem findings

It may interest veterinary student readers to know that I have found over the years that the presence of fluid (usually blood-stained) in the pericardial sac (*photo 4*) (i.e. the sac around the heart) is diagnostic of enterotoxaemia. The fact allows a fairly rapid confirmatory field diagnosis.

Prevention

There is no excuse for any flockmaster who loses lambs or sheep from enterotoxaemic diseases. If used correctly, the vaccines and antisera available are almost 100 per cent efficient. However, for some reason which I shall never be able to understand, sheep farmers again and again forget to, or do not vaccinate.

When an outbreak on such farms does occur, move the remaining sheep on to a bare pasture immediately or restrict their diet for several days before gradually building them back up to the original plane of nutrition.

In the meantime, having confirmed the diagnosis by consultation with the veterinary surgeon, protect the remaining flock with antisera pending the introduction of a full vaccination programme.

The compound vaccines are the best and most economical. Your veterinary surgeon will advise on the correct one to use to cover your particular disease problems.

Just one important point – booster vaccination immediately before lambing will give the lambs, via the colostrum, a protection for the first 12 weeks of life. Also a second booster two weeks before flushing will protect against the improved feeding. If you are able to market your lambs by that time, then there is no danger. However, all lambs kept over 3 months, whether as replacement breeders or fatteners, *must* be vaccinated. The ideal is to give them their first dose at 6 weeks and their second at 12 weeks. The latest vaccines also protect against pasteurellosis, and some combine anthelmintic treatment.

12 Lamb Dysentery

This is without doubt the most dangerous of all sheep diseases. It attacks lambs under a week old, though just occasionally it can knock them down at 2 or 3 weeks.

Cause

It is caused by *Clostridium perfringens* Type B. This germ produces its effect in the same way as Types D and C, that is, by multiplying in the small intestine of the lamb producing haemorrhages and excreting highly lethal toxins (*photo 1*). The mortality rate among infected lambs is virtually 100 per cent.

It is a particular problem in the border counties between England and Scotland and in North Wales, but it can flare up anywhere.

Symptoms

The first sign noticed is usually the sudden death of one or two lambs. Closer observation will show that several others are dull and lethargic. They stop sucking and may start to kick at their belly. They stand with their backs arched and when made to move, they stagger rather than walk (*photo 2*), then flop down. Within 24 hours, if still alive, they develop a profuse brown, often bloodstained diarrhoea (*photo 3*). In the words of one of my shepherd friends, 'The poor little beggars strain their guts out'. They become dehydrated, comatose, and die approximately 1 day after the onset of the diarrhoea.

In 2- to 3-week-old lambs the diarrhoea is less severe and the patients may survive for 2 or 3 days.

Treatment

Once the symptoms have developed, treatment is a waste of money and time, but each of the unaffected lambs should be protected immediately by injecting them with 2cc of concentrated lamb dysentery serum. Any further new-born lambs during that lambing season must be given the serum as soon after birth as possible.

Prevention

Obviously this disease must be controlled, and it can be by vaccinating the ewes in all subsequent years with a multivalent sheep vaccine. Two doses should be given the first year – at 6 weeks and immediately before lambing – and a booster dose immediately prior to each subsequent lambing.

Any new breeding stock must have the double vaccination during the first year on the farm.

Again, as with enterotoxaemia, a second booster to the ewes two weeks before flushing will reinforce the potential immunity of the lamb crop.

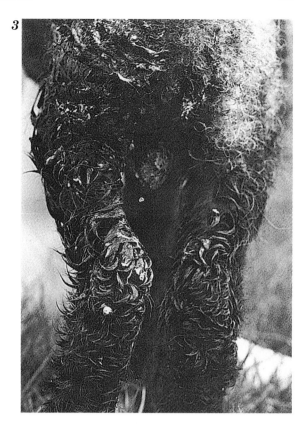

13 Braxy

This is a very acute disease of young sheep occurring mainly in hill flocks between October and March.

Cause

It is caused by *Clostridium septicum* (*see diagram*) which also excretes lethal toxins; the growth of this germ occurs in the ewe's fourth stomach (the abomasum).

The *Clostridium septicum* multiplies in the wall of the abomasum producing a powerful toxin, which is absorbed and causes a toxaemia.

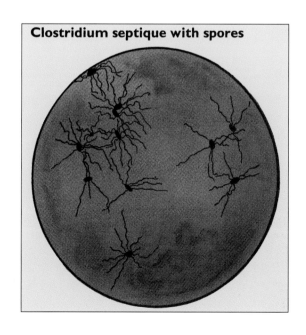

Clostridium septique with spores

Outbreaks usually coincide with the onset of frost.

Symptoms
It is usual to find several sheep dead. If seen alive, the braxy patient stands away from the rest of the flock. It is dull, refuses to eat, runs a high temperature of around 107°F (41.6°C) and shows signs of abdominal pain, viz. grunting, grinding of the teeth and shifting the feet.

Within an hour or two it lies down; the temperature drops to below normal and death follows rapidly.

Post-mortem signs
If the carcase can be opened up immediately after death, there is a clearly marked area of purple congestion and/or ulceration on the lining of the fourth stomach (*photo 1*). Unfortunately this area disappears if the carcase is not examined within a few hours of death.

Treatment
There is none.

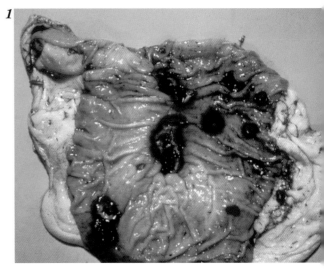

Prevention
Comprehensive vaccination, in all braxy areas, using an appropriate multivalent vaccine twice the first year and once each subsequent year during the early part of September.

14 Black Disease

An acute fatal toxaemic disease affecting mainly adult sheep (*photo 1*).

Cause
Another member of the Clostridial family – *Clostridium novyi* Type B, which multiplies in damaged liver tissue.

Mostly the trigger factor is fluke infestation. The liver damage that is caused by the invasion of immature flukes presents an ideal medium for the germ's growth (*photo 2*). The bug proliferates rapidly and excretes lethal toxins that are then absorbed into the bloodstream.

When and where it occurs

It is found on any hill or lowland farm that is infected by *Fasciola hepatica* (the sheep fluke).

Outbreaks coincide with the activity of the fluke parasite, that is, chiefly in the autumn and early winter and just occasionally in early spring.

Adult sheep are the more susceptible, though I have seen outbreaks in lambs facing their first winter.

Symptoms

Again, comparatively sudden death is the rule since the course of the disease is a short one.

If seen in the early stages, the affected sheep lag behind the rest of the flock. When driven they collapse and breathe heavily. If left alone, they appear to fall asleep lying on their belly and brisket with their head round to one side (*photo 3*). Any sudden noise will make them start violently and the eyes and ears twitch nervously. Shortly afterwards they go into a coma and die within a few hours.

Treatment

Treatment is useless in affected animals, but the remainder of the flock should be protected immediately with black disease antiserum. This gives only up to 20 days' protection, so it is wise to give a first dose of black disease vaccine at the same time as the serum, and then to give a second booster dose a month later.

Prevention

Obviously this is another disease which should never be allowed to flare up. All sheep being kept in fluke areas must be protected by annual vaccination with a multivalent vaccine – twice the first year and a single booster each subsequent August or September. In addition, of course, appropriate measures should be taken against the fluke (see page 72). Again, a second booster before flushing will increase the immunity. An oily intra-peritoneal vaccine is also available.

15 Clostridial Infection of Wounds

This general heading incorporates three commonly recognized conditions:

Post-parturient gas gangrene.
Blackquarter – blackleg (photo 1).
Malignant oedema (diagram A).

Cause

The clostridia involved in post-parturient gas gangrene and in other wound infections are:

Clostridium septicum (malignant oedema);
Clostridium perfringens;
Clostridium chauvoei;
Clostridium novyi;
Clostridium sordelii.

In blackquarter or blackleg, *Clostridium chauvoei* is the bacterium most frequently involved (*diagram B*). In so-called malignant oedema, the chief 'offender' is the *Clostridium septicum*.

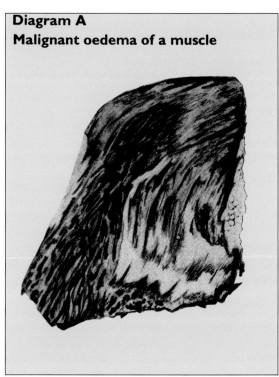

**Diagram A
Malignant oedema of a muscle**

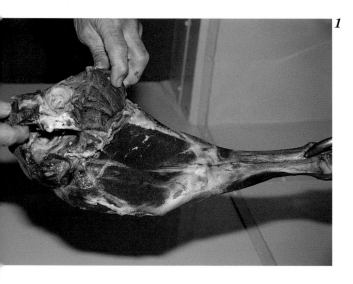

1

When infection is likely to occur

Obviously, infection is most likely when wounds are inflicted on the sheep: for example, during or after lambing, castration, docking or shearing. Head wounds caused by rams fighting are also susceptible.

Deep puncture wounds are the most dangerous. Unlike the blackleg of cattle, blackquarter in sheep flares up only as a result of wound infection.

Symptoms

Two days after the respective wounds become infected the sheep become depressed and dull

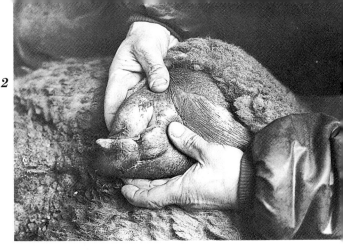

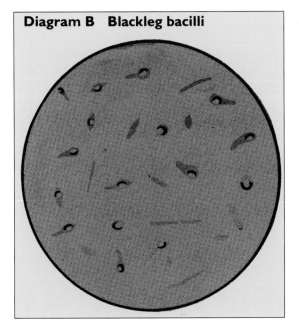

Diagram B   Blackleg bacilli

and run a very high temperature (107°F, 41.6°C). They walk with difficulty, the wound area is swollen (*photo 2*), soft and hot but later becomes tense and hard. Pressure over the area usually shows crepitation, that is, when you run your fingers over the skin surface, it feels as though there is tissue paper underneath: although this may not be present when only *Cl. novyi* is involved.

Treatment
The only hope is to catch the condition very early – then daily injections of penicillin or broad-spectrum antibiotics may possibly save the sheep. Usually, however, cases are too far gone before being seen. Obviously, therefore, it is much better to do everything possible to prevent such infections.

Prevention
All these conditions can be controlled by the routine use of a comprehensive vaccine covering the offending clostridia. Give two doses of the multivalent vaccine the first year, then a booster immediately before lambing, and another

booster each year, again just before lambing.

Despite this protection, it is always wise to use strict hygiene precautions at lambing time.

Also, a routine which I invariably advise after every bad lambing case – in fact whenever any assistance has to be given – is a single dose of long-acting penicillin injected, preferably, intramuscularly (*photo 3*).

16 Tetanus

Tetanus is primarily a disease of lambs, though all ages of sheep can be affected.

In the lambs it occurs 2 to 3 weeks after castration and docking, especially where the rubber ring method is used (*photo 1*).

In ewes it may be seen shortly after lambing or shearing.

As a rule, several animals are affected and, like braxy and black disease, tetanus tends to occur on the same farm year after year.

Cause

It is caused by the toxins produced by the germ *Clostridium* or *Bacillus tetani* (*see diagram*), which is commonly present in animal dung and in the soil.

Clostridium tetani is a sporulating bacterium, i.e. it surrounds itself by a protective covering which enables it to live in the soil for many years.

Any wound, particularly behind or underneath the animal, is liable to become contaminated by the tetanus spores.

If the wound is deep, the bug casts off its

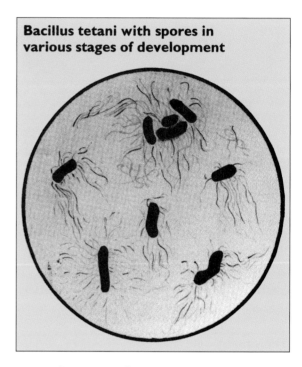

Bacillus tetani with spores in various stages of development

protective coat and starts to multiply.

Two toxins are excreted and one of these – a neuro (or nerve) toxin – travels along the nerve fibres to the central nervous system.

The damage it causes in the central nervous system (the brain and spinal cord) results in increased irritability and acutely painful muscle spasms.

Symptoms

The symptoms may be mild or acute.

In mild cases the affected animal merely appears stiff and walks in a peculiar stilted fashion (*photo 2*), just as though it were suffering from the after-effects of the castration, docking or lambing. Such mild cases

1

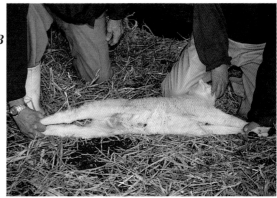

may recover completely without treatment.

In acute cases the stiffness is much more obvious and widespread. The muscles are hard and rigid, the head and neck are extended, and the tail cocked.

The patient is hypersensitive; it has great difficulty in lowering its head, and the jaws become locked. Eventually it falls over on its side and lies with legs outstretched and head thrown back (*photo 3*).

Treatment
Mild cases appear to respond to daily injections of penicillin (6cc of 300,000 units per cc).

Acute cases do not recover and should be destroyed.

Prevention
Comprehensive vaccination, coupled with strict aseptic precautions and general cleanliness during castration, docking, lambing and shearing, i.e. whenever a wound is likely to be produced.

With valuable pedigree lambs an immediate booster protection can be given by injecting with tetanus antitoxin, which reinforces the immunity for a month.

17 Botulism

This is an *uncommon* disease of sheep seen mainly in starving and scavenging animals, though it has been seen in sheep fed big bale or badly made silage.

Cause
The toxin of *Clostridium botulinum*.

Source
Decomposing animal material. The starving scavengers are so short of protein and phosphorus that they will literally eat anything. When they swallow the *Clostridium botulinum*, the lethal toxin causes a progressive muscular paralysis.

Symptoms

The affected sheep salivates profusely (*photo 1*) and walks stiffly. Within a few hours it starts to bob and sway like a drunken person and has great difficulty in breathing. Eventually it flops down and dies, simply because the chest muscles become paralysed and it can no longer draw breath.

Treatment

There is none.

Prevention

Reasonable feeding – good hay is preferable to silage. *Never starve your sheep.*

To the average sheep farmer the inclusion of this disease may seem academic, but it is wise to be constantly aware of the dangers of malnutrition, particularly on remote hill sheep farms.

Diseases of Tick Areas

18 Louping Ill (Trembling)

This is a virus disease spread by ticks and affecting hill sheep chiefly in Scotland and northern England. It breaks out during the months of greatest tick activity – April, May, early June and September.

When it appears in a flock for the first time, it affects sheep of all ages and can kill up to 60 per cent. Once it is established, it recurs each year but much less virulently, attacking chiefly the yearlings and killing no more than 4 per cent of the flock. Recovered animals have a lifelong immunity.

Fortunately not all tick areas have the louping ill virus. It can occur in other areas of England and Ireland, and an identical condition has been described in Russia and Europe.

Symptoms

After an incubation period of 1 to 3 weeks, the virus invades the bloodstream causing a high fever. The affected animals are dull, they breathe heavily, are usually off their feed (*photo 1*), and have a temperature up to 106°F (41°C).

The fever subsides after a few days and many recover, but a week later typical nervous symptoms may develop. The patient becomes hyperexcitable and starts to tremble (hence the name, 'trembling'). The trembling

is most marked around the head and neck. At this stage the temperature may rise again.

As the disease progresses, the patient often appears to be blind. It may start to prance like a trotting pony or it may suddenly leap forwards and fall flat, kicking its legs as though in a fit. Coma and death follow.

Treatment

There is no satisfactory treatment: sedation by heavy doses of tranquilliser plus hope. If paralysis of the hind legs does not occur, the patient stands a chance of apparent recovery, though it may remain a carrier of the virus despite having a lifelong immunity.

Prevention

Obviously it is vital to control the tick so far as possible. Regular dipping combined with heather burning and bracken cutting will assist greatly.

However, in all tick areas it is wise to combine such precautions with the rigid routine use of the louping ill vaccine. All sheep should be vaccinated twice the first year, in the early spring (March) and late summer (August), and thereafter, if the flock remains self-contained, all ewe lambs should be treated similarly and all breeders should be given a booster dose at least every other year before lambing.

Just one more interesting point. Louping ill is transmissible to man, causing a high fever and acute muscular pain, though fortunately it is rarely fatal in humans. Recently cases have been reported in working dogs, cattle, pigs and horses.

General control measures against ticks and tick-borne disease

1. Burn the heather, bracken and rushes.
2. Plough and reseed wherever possible.
3. Dip the ewes after lambing and twice thereafter at 6–8 week intervals.
4. Dip the lambs and repeat twice at 14-day intervals.
5. Wherever possible keep the flocks self-contained.
6. Vaccinate against louping ill as directed.

19 Tick-Borne Fever

A non-fatal but debilitating fever which can attack sheep of all ages in the tick-infested areas of Scotland and northern England and in Scandinavia. Like louping ill and tick pyaemia, it occurs when ticks are most active in April, May and early June, and again in September. In a tick-infested area probably every lamb contracts it within the first 2 weeks of life. Cattle are also susceptible.

Cause

A rickettsial organism which is transmitted by the tick *Ixodes ricinus* (*see diagram*). The source of infection is the blood of an infected sheep or of a carrier sheep, or carrier bovine.

Symptoms

Adult ewes may abort and rams may become infertile. Affected animals run a temperature of between 104° and 107°F (40° and 41.6°C) for 9 or 10 days. The fever then subsides, but

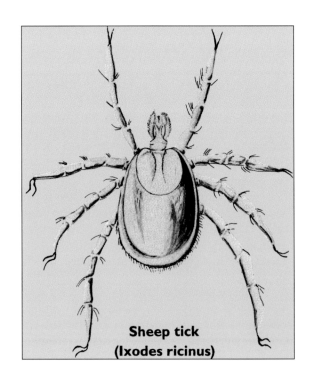

**Sheep tick
(Ixodes ricinus)**

not before it has played havoc with the animal's condition and has lowered resistance to diseases such as louping ill or tick pyaemia. In this respect it is not unlike the glandular fever seen in humans. After the fever there is a dullness and inappetance for 7 days.

Fortunately the infertility in rams lasts for only approximately 3 months.

Lambs become infected very early in life, usually showing only mild symptoms such as dullness and failure to suck. They then develop an immunity.

Treatment
Long-acting tetracyclines appear to stop the disease within 24 hours. Also a 3-day course of sulphadimidine is effective.

Prevention
Maximum control of the ticks by regular dipping, bracken cutting and heather burning is, so far, the only defence against tick-borne fever, as the scientists have not yet succeeded in producing an efficient vaccine.

A commonsense management precaution is never to move non-immune bought-in ewes into a tick area during pregnancy.

20 Tick Pyaemia (Enzootic Staphylococcal Infection)

This is a blood poisoning of lambs occurring in tick areas in April and May when the ticks are most active (*photo 1*). It affects lambs between 2 and 16 weeks and the majority at 3 to 5 weeks of age.

1

Cause
The same germ that causes most cases of mastitis in ewes, namely, *Staphylococcus aureus*. It gains entrance to the body through the tick bites, and if the lambs live long enough, it produces multiple abscesses throughout the body including joints, tendon sheaths, ribs, spinal column (*photo 2, overleaf*), brain, liver, spleen, kidneys, heart wall and lungs.

Symptoms
In hyperacute cases, sudden death occurs.

In the less acute cases, the affected animals run a temperature of between 104° and 107°F (40° and 41.6°C) (*photo 3, overleaf*) for 9 or 10 days. Thereafter the typical symptoms take longer to manifest themselves and vary according to where the abscesses form. For example if the abscess is in the spine (as illustrated in *photo 2, overleaf*)

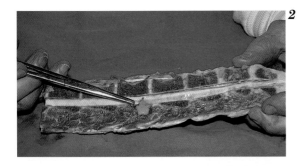

paralysis of the hind legs may develop; if in the brain, signs of meningitis; if in the spleen, heart, liver, kidneys or lungs, progressive unthriftiness; if in the tendon sheaths or joints, marked lameness. In fact more than half the lamb losses on a tick farm and nearly all lamenesses are due to tick pyaemia.

Treatment

Once affected, most lambs die or have to be destroyed. Mortality may be up to 20 per cent of the lamb flock.

If seen very early, a 5-day course of procaine penicillin (300,000 units daily) injections or a single injection of long-acting tetracycline might effect a cure. Advanced cases are much better destroyed.

Prevention

Again, commonsense control of the ticks is vital but here the obvious snag is that the lambs will not have been dipped.

Where tick pyaemia is a problem, the lambs should be dipped shortly after birth. Because of a lamb's short coat this may only protect for 14 days so they should be dipped again before being turned out on the hill.

There is no satisfactory vaccine, and routine preventative use of antibiotic injections would be impracticable.

Additional Diseases of Lambs

21 Scour in Lambs (Neonatal Diarrhoea)

E. coli infection

Since the nightmare of lamb dysentery has been more or less completely subdued by general vaccination, another form of scour in lambs (*photo 1*) has moved in to take its place as the chief menace to lambs during the first

1

week of life. This is *Escherichia coli* infection. Rotavirus and cryptosporidium may also be involved, though the cryptosporidium is the active menace in lambs 2–3 weeks old.

Cause
A bacterium called *Escherichia coli*. There are several different strains of this germ.

The predisposing cause is lambing in a confined area. *E. coli* infection is usually most severe towards the end of lambing, indicating a build-up of infection.

Symptoms
The affected lambs stop sucking and stand around with their backs arched (*photo 2*).

2

They develop a bright yellow diarrhoea and, if untreated, they die within a couple of days, usually from dehydration.

A peculiar feature of the disease is that it does not always seem to be acutely contagious, because one lamb of twins or triplets may be affected while the others remain perfectly normal.

Some of the affected lambs may become hyperexcitable and even blind (i.e. when a septicaemia occurs).

Treatment

Immediately isolate those affected. Get your veterinary surgeon to take swabs of the faeces to have the *E. coli* typed and a sensitivity test done. He will provide an oral general antibiotic mixture (*photo 3*) as immediate protection pending the laboratory report and will probably provide oral electrolytes to counteract the dehydration. When the disease is confirmed, it will

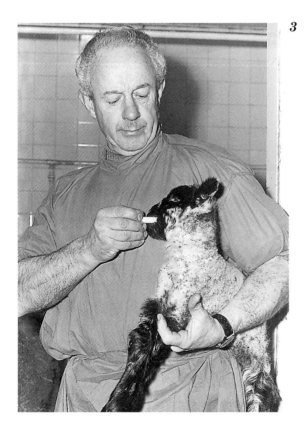

3

probably be necessary to dose all new-born lambs with the specific antibiotic.

There is no specific treatment against rotavirus or cryptosporidium. Nursing is the major factor.

Prevention

Move the ewes on to a fresh lambing area half-way through lambing or, better still, move them twice during the lambing period.

Occasionally, the *E. coli* strikes again when the lambs are 2 to 4 weeks old, though at this stage the main menace is the cryptosporidium. The stress factor here is usually overfeeding the ewes with high protein food combined with a too generous creep feed for the lambs.

The symptoms and treatment are identical to those described for the younger lambs. Once again, a laboratory sensitivity test is vital to find the appropriate antibiotic. However, the outbreak can usually be terminated by halving the feed or substituting a coarse grain ration for the high protein nuts and pellets.

E. coli vaccines are available for immunizing ewes. Their effect is variable but they are well worth routine use.

Watery mouth

Seen frequently in lambs under a week old, and most often in triplets or lambs out of poor ewes.

Cause

Constipation associated with retention of the meconium (or first dung), due usually to lack of colostrum.

Symptoms

Salivation, often profuse, causing drooling at the mouth and soiling under the chin (*photo 4*). The lambs stop sucking and die rapidly if untreated. When you shake the lamb you may hear splashing or gurgling sounds. Death is due to an *E. coli* septicaemia.

Treatment
Oral purgatives plus liquid paraffin enemas with a 3- or 4-day cover of antibiotics to prevent the septicaemia. Start the antibiotics at the slightest early symptom.

Prevention
As with *E. coli* build-up – commonsense plus good hygiene. Also make sure the new-born lambs get their full quota of colostrum (i.e. the ewe's first milk).

Do not apply rubber rings at birth, wait for 24 hours at least. It is well worth trying an *E. coli* vaccine at birth, plus the oral antibiotic.

Salmonellosis

Salmonellosis occurs only sporadically in sheep and then usually under housed conditions. Needless to say the lambs are the most acutely susceptible (*photo 5*).

Cause
Three types of salmonella, namely *S. typhimurium*, *S. dublin* and *S. montevideo*, the latter being much less common.

Symptoms
The lambs stop sucking and may run a high persistent fever. They are dull and depressed, with diarrhoea which may or may not contain blood. Death occurs from septicaemia.

When a ewe becomes infected, the symptoms are even more severe (*photo 6*).

Treatment
Much more difficult to treat than *E. coli*, with antibiotics often having little or no effect.

In most cases, isolation and careful nursing plus oral electrolytes (to avoid dehydration) provide the only chance of survival.

Prevention
Salmonella vaccination of the ewes is unsatisfactory so prevention should be aimed at, namely repeated movement of the lambing ewes plus hygiene and cleanliness. Also, suspected cases should be isolated immediately.

Coccidiosis

This disease can produce diarrhoea in lambs 1 to 4 months old. The general picture is that the majority of outbreaks occur in lambs 4 to 8 weeks old that were housed at birth. It also occurs in 6-week-old lambs that were turned out 2 weeks previously. It is much less common in lambs that have not been housed.

The ewes are the source of infection. Although they have a natural immunity they contaminate the bedding with the coccidia oocysts and the lambs pick them up almost immediately. Fortunately the lambs start developing an immunity after 6 weeks but in the meantime the oocysts produce the typical signs of the disease.

Cause

Two species of coccidia called *Eimeria ovinoidalis* and *Eimeria crandallis*.

Symptoms

The usual sign is dark, bloodstained diarrhoea which soils the hind-quarters (*photos 7 & 8*). If the lambs are in good condition, they recover fairly rapidly without treatment.

In other cases, however, the bloodstained diarrhoea persists and the lambs become tucked up and unthrifty if the correct treatment is not prescribed. In fact, untreated cases can result in death from dehydration.

Treatment

Call in your veterinary surgeon. He will take dung samples to identify and confirm the causal parasites and prescribe the appropriate specific treatment.

One of the best-tried methods of treatment has been a 5-day course of an anticoccidial agent by the mouth, followed by a complete change of pasture.

Prevention

One ideal is to provide slatted floors in the lambing pens. Failing that, remove all damp and dirty bedding daily and give a preventative dose of the anticoccidial drug at 3 and 6 weeks.

Move the lambs to fresh and extensive grazing as soon as possible.

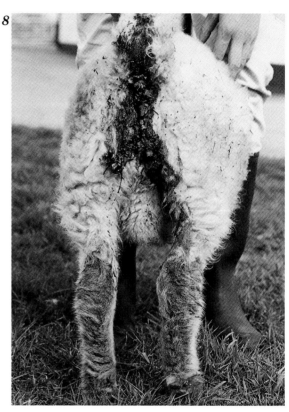

From 6 to 12 weeks of age, nematodirus infestation is the commonest cause of scour in lambs (see Nematodirus Infection, p. 66). Thereafter, ordinary stomach and bowel worms are the main culprits (*see* Roundworms, p. 62).

22 Daft Lamb Disease (Cerebellar Atrophy)

Like swayback, daft lamb disease is a condition of new-born lambs and is the only disease I know that is likely to be mistaken for swayback; in fact the two conditions are sometimes indistinguishable (*photo 1*). Fortunately, however, daft lamb disease is uncommon and is seldom, if ever, a major problem. It is seen chiefly in the Border Leicester breed and its crosses, e.g. the Scottish halfbred and the Greyface.

Cause
Daft lamb disease is *not* due to a mineral deficiency but is genetic in origin. Part of the lamb's brain – the cerebellum – is congenitally atrophied (imperfectly developed).

Symptoms
Most cases have a characteristic jerking backwards of the head. They carry the head high and have the mouth pointing backwards

(*photo 2*) or towards the side. In the more severe cases the lambs are blind and walk in circles before flopping down (*photo 3*). Despite this the appetite remains good.

Treatment
There is none, though mild cases can be hand reared and should be given a chance.

Most cases are destroyed by shepherds or

1

2

3

die of exposure or starvation. Those that can walk may improve gradually, and some may become apparently normal except when excited.

Prevention

Cull the mothers and change the tup. One obvious preventative measure is to avoid buying rams from flocks where the condition is known to exist.

23 Joint-ill (Bacterial Polyarthritis)

A disease affecting the joints of young lambs up to one month old and seen especially when the ewes are lambed in yards (*photo 1*) or in dirty pens in a field. In such conditions the disease incidence can be very high.

Cause

Various organisms, chiefly *E. coli* and several streptococci, that gain entry to the blood-stream via the umbilicus or navel cord shortly after birth. Staphylococci and pasteurella are often involved. These organisms may also gain entry through castration or docking

wounds; or through tick bites. *Coryne-bacterium ovis* may be a causal agent or a contaminate but not, so far, in the UK.

Symptoms

An initial fever with the temperature rising to 105°–106°F (40.5°–41°C) stops the lamb sucking. It becomes tucked up and then suddenly becomes lame with one or more joints markedly swollen and painful (*photo 2*). Occasionally the swelling extends down the

2

1

tendons at the back of the affected leg. The shoulder, elbow and stifle joints seem most prone.

3

Treatment

If caught early, i.e. as soon as they stop sucking and show the high temperature (*photo 3*), most cases respond to daily injections (for 5 days) of penicillin and streptomycin, or to long- or short-acting broad-spectrum antibiotics such as tetracycline. However, once the joints become infected the majority of the lambs die within a week, while the few that recover do so slowly and remain lame and unthrifty.

Prevention

Obviously this is a disease that shepherds should do everything possible to avoid, particularly since there is no satisfactory vaccine against it.

The accent must be on hygiene and general cleanliness.

Lambing pens should be bedded down with clean straw between each patient and the lambs' navels should be dressed as soon after birth as possible (*photo 4*). The most convenient and perhaps the best dressing is an aerosol spray containing a broad-spectrum antibiotic. If this is not available, tincture of iodine is an excellent substitute.

Shepherds should use abundant water, soap, soap-flakes and antiseptic whenever assistance is necessary, and should dip their feet in antiseptic solution before entering the lambing pens. At castration and docking, clean (preferably sterile) instruments should always be used.

Also the lambing paddock should be moved to clean ground at least once and preferably twice during the lambing season.

Certainly joint-ill can be controlled if

4

conditions of hygiene are maintained, not only at lambing time but during the castration and docking procedures.

It has been my experience that since the general introduction of rubber ring castration and docking, joint-ill cases from these sources are much fewer than tetanus cases.

24 Liver Abscesses (Hepatic Necrobacillosis – *Fusiformis Necrophorus* Infection)

This condition can occur at any time on most lowland farms. It affects lambs 5 to 14 days old (*photo 1*) and can kill off 5 to 10 per cent of the lamb flock. The odd case may be seen among hill flocks.

Cause

It is caused by a germ called the *Fusiformis necrophorus* – the same germ that is involved in foot rot (*see* p. 73). The *Fusiformis* gains entrance through the navel in new-born lambs.

Symptoms

One or several lambs that have been doing really well for the first few days suddenly stop sucking and literally fade away. The abdomen often swells up and death follows in a few days.

The disease is confirmed by post-mortem examination, when numerous white abscesses (*photo 2*) or caseating, vile-smelling dead nodules are found in the liver and occasionally in the lungs also.

Treatment

As in joint-ill, treatment has to be started *very* early, i.e. just as soon as the lamb stops sucking. Then a 3-day course of penicillin injections (*photo 3*) will often save the patient

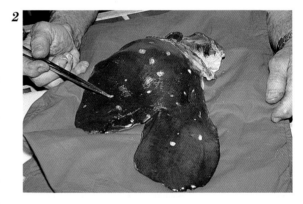

2

1

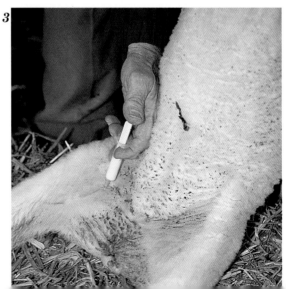

3

(approximately 300,000 units of penicillin per day). Long-acting tetracycline may also be effective, though most often the cases treated show only a temporary improvement.

Obviously, many cases will be too far gone before treatment can be applied, so it is better to avoid the danger of infection.

Prevention
The advent of the foot rot vaccine should do much to control this condition. The entire ewe flock should be fully protected at all times. In this way the danger of the lambing pens becoming contaminated will be reduced to a minimum.

The same hygienic precautions as advised for joint-ill should be strictly imposed, including the spraying of the navels of the new-born lambs with a broad-spectrum antibiotic, or dressing with tincture of iodine.

Where the lambing is being done in the open pasture, a change of field will often prevent new cases developing. By the same token indoor lambing pens should be moved to a clean area at least once during the lambing season.

Note: The scientists have altered the name of the causal organism to *Fusobacterium necrophorum* – personally I prefer the original: the name change is purely academic.

25 Stiff Lamb Disease

A disease of growing lambs, 3 to 4 months old. In Britain the highest incidence occurs in East Anglia on arable land.

Cause
A germ called *Erysipelothrix rhusiopathiae*, i.e. the same germ that causes swine erysipelas.

The *Erysipelothrix* bug is believed to come from the soil and to get into the lambs through docking or castration or other wounds. It may also be eaten by the lambs (*photo 1*).

Rarely do more than 5 per cent of an infected flock show signs of the disease, the onset of which is gradual. But just occasionally up to 25 per cent may show the tell-tale stiffness.

What usually happens is that one or several lambs may be seen to be slightly lame; later a few more may be noticed. Rarely is there any history of contact with pigs.

Symptoms
The affected lamb is lame, but there is usually no obvious swelling of the joints. The lamb merely walks stiffly – hence the name, 'stiff lamb disease' (*photo 2*).

The lameness persists for a long time and the patients do not thrive, simply because the *Erysipelothrix rhusiopathiae* produces a painful deep-seated fibrinous arthritis (i.e. an arthritis with no pus) (*photo 3*).

Treatment
In early cases, penicillin injections will destroy the bugs and prevent a chronic arthritis. Once the condition has become well established, however, treatment has little chance of success. Obviously, therefore, as with so many of the other sheep diseases, it is much better to do everything possible to avoid stiff lamb disease.

Prevention
Strict asepsis at docking and castration. The lambs should be dropped on a pile of clean straw and then turned on to a fresh pasture immediately after the operation.

They should be kept at grass until the wounds are healed. In other words, ***never turn newly docked or castrated lambs on to arable land***.

I have used swine erysipelas vaccine with some success, injecting the ewes with 2cc at the same time as the multivalent clostridial vaccines. But the disease can and should be controlled by simple cleanliness and common-sense.

The advent of rubber ring castration and docking has undoubtedly reduced the incidence.

An acute condition does occasionally occur but most often the disease manifests itself in the chronic form. When acute the lamb will be dull and off its food before the stiffness develops.

Occasionally stiff lamb disease appears after dipping when it has been called 'Post-dipping lameness'. Such cases are put down to the dip not being freshly prepared or not containing an effective bacteriostat (i.e. a drug to stop bacteria multiplying).

A new vaccine called Erysorb ST provides improved protection against erysipelas in lambs, ewes and rams (*photo 4*). With this new vaccine the very young lambs receive immediate immunity from the mother's colostrum. To achieve this the vaccine should be given to the ewe twice in 2ml doses, 2–6 weeks apart; with the second dose at least 4 weeks before lambing. Subsequently a single annual booster dose is required.

3

4

26 Hypothermia

Without doubt the greatest cause of losses in lambs in most sheep countries of the world is hypothermia: that is, rapid loss of body heat or, to put it even more simply, sudden cold immediately after birth, often coupled with starvation. Twins and triplets are at greatest risk, along with lambs from old ewes or, in some instances, from very young mothers.

In recent years considerable scientific research has been done on hypothermia, details of which are mainly for the scientist. However, the findings are of vital importance to the practical sheep farmer.

More and more it is becoming apparent that by far the best way to combat hypothermia in countries with a severe or even a moderately severe winter is to adopt a system of *controlled* indoor lambing.

In my experience in the Midlands of England the ideal is to bring the entire ewe flock indoors into well-ventilated but draught-proof buildings a *fortnight* before lambing starts.

As each ewe lambs, place the mother and her progeny in a pen with ad lib hay and water plus a trough for concentrate feeding. Keep each family in the pen for approximately one week; then when the lambs are fit and strong, turn them out onto the pasture with the mother (usually when the lambs are two weeks old).

This method not only avoids lamb losses from hypothermia but makes the shepherd's job infinitely easier. It also allows much prompter attention to conditions such as calcium and magnesium deficiency, mastitis, scour in the lambs, starvation, etc.

Another considerable economic advantage is that it can provide a second crop of wool if the ewes are shorn when brought inside.

Provided that hygiene is strictly observed, and that none of the flock is kept housed for excessively prolonged periods (6 to 8 weeks is the average), the dangers of constant indoor management are avoided (see Protecting the Indoor Ewe and Lamb, p. 173).

The series of photographs that follow on pages 58 and 59 illustrate the advantages of a highly successful set-up with which I have been involved. It is based largely on experience and commonsense, and is managed under ordinary farm conditions using old buildings.

Obviously there will be variations on the various farms depending on the buildings and labour available, but the general procedure can be followed by any sheep farmer with the utmost confidence.

During indoor wintering, hay is essential as is fresh, clean drinking water.

A good compound mineral containing traces of iron can be dispersed in troughs and is ingested during the eating of the concentrate rations. The nut concentrate should contain added magnesium to prevent hypomagnesaemia.

Each pen has fresh water twice daily in an improvised plastic container and another improvised plastic trough for feeding the concentrates.

Improvised individual lambing pens with hay racks and fresh straw bedding daily. The floor area is scrubbed out with hot water and soda between each ewe to prevent joint-ill and E. coli infection.

Suckling the weakly lamb in an individual pen is much easier. Also it is much easier to judge whether a lamb is getting enough milk. It is also easier to inject day-old lambs if this is necessary.

Indoor lambing appears to produce stronger and more vigorous lambs. The lambs shown above had been turned out for only 3 weeks and were well on the way to weaning.

By the same token, the shepherd can make sure the lamb gets colostrum from dropped udders. Cow's or goat's colostrum can be given by stomach tube if necessary (see page 60–1).

The successful rearing of quads, made possible only by indoor lambing.

Lambs ready for turning out, 10–14 days old.

Practical control of hypothermia

As already mentioned, hypothermia can be due to:

- *Exposure* to inclement weather conditions at or immediately after birth. This can cause the lamb's body temperature to drop as much as 1°C (2°F) per minute, with death a near certainty if prompt treatment is not given.
- *Starvation* – seen in slightly older lambs varying in age from approximately 8 to 80 hours. This type of hypothermia can occur regardless of environmental conditions and is usually due simply to the lamb having

stopped sucking. Once again there is a fall in body temperature, but in addition a low level of blood sugar quickly manifests itself and it is this shortage of sugar (hypoglycaemia) which causes death.

- *Disease* such as scour, watery mouth, pneumonia etc. Whereas housing at lambing time combined with careful feeding and keen observation can do a great deal to prevent exposure and starvation, the disease cases require specialized veterinary attention.

How to treat
These are the tools for the job, as prescribed and supplied by your veterinary surgeon:

- six 50ml plastic syringes
- two stomach tubes
- one or two bottles of a sterile 20 per cent glucose solution
- powerful skin antiseptic
- two 10ml syringes and several sterile fine 25mm (1in) needles
- a bottle of long-acting antibiotic

In addition you will require a sterile jug, two or three dry towels, and a cosy pen deeply bedded in straw or preferably a commercial warming box.

Technique
Simple hypothermia with the lamb under 4 hours of age
1. Dry the patient thoroughly with a warm towel.
2. Milk off colostrum from the mother ewe into the sterile jug.
3. Provided the lamb can hold its head up, pass the stomach tube into and down the lamb's oesophagus (*photo 1*). This is remarkably easy but it is wise to get your veterinary surgeon to teach you the technique.
4. Draw 50ml of the colostrum into one of the large syringes and attach the syringe to the stomach tube. Insert the plunger slowly (*photo 2*). Refill the syringe and repeat the dose until at least 200-250ml of the colostrum has been given.

5. Inject 2ml of long-acting antibiotic into a shoulder muscle.
6. Wrap the lamb in a warm piece of blanket, or ideally place it in a commercial warming box.
7. Repeat the feed every 8 hours until the lamb is on its feet and able to suck the ewe.

Hypothermia due to starvation (usually when the patient is over 8 hours old)
The treatment is identical to the above but in addition an injection of the 20 per cent solution of glucose is absolutely vital to the lamb's survival.

1

2

If, when spotted, the starving lamb cannot hold its head up, then get your veterinary surgeon to inject the glucose immediately and don't attempt to pass the stomach tube until it can raise the head.

The required dose of the 20 per cent glucose solution is between 30 and 50ml depending on the size of the lamb. The glucose solution has to be warmed to blood heat and given intra-peritoneally using strict aseptic precautions (*photo 3*); hence my advice to use a veterinary surgeon initially, at least until the shepherd has mastered the correct technique.

Any unused colostrum can be stored in a refrigerator to be used as a back-up in case of subsequent problems.

Internal Parasites

27 Roundworms (Gastro-intestinal Helminthiasis) (Nematodes)

Roundworms in sheep (*photo 1*) still produce greater losses to the sheep industry than any other single cause or disease, despite the free availability of the many excellent modern anthelmintics.

In my opinion this is due to haphazard dosing, combined with the intensification of stocking associated with improved grassland management; and also the development of worms resistant to certain of the worm remedies. Productive pasture leads to heavy stocking and heavy stocking leads to excess parasites.

In order to understand this fully and to appreciate the necessity of good pasture husbandry and regular routine dosing, a knowledge of the life cycle of the worms is absolutely vital.

Life cycle of the sheep roundworms

Approximately 10 different species of roundworm live in the stomach and intestines of British sheep, and all of them basically have the same life cycle (*see diagram*).

Eggs laid by the adult female worms inside the stomach or bowels, are passed out with the sheep's droppings on to the pasture.

In 24 hours the eggs change into first-stage larvae which feed on bacteria and moisture.

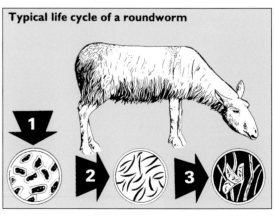

Typical life cycle of a roundworm

1

Within a short time the first-stage larvae change into second-stage larvae. These also feed on bacteria and moisture, then form the third or infective stage. The infective stage retains the protective skin of the second stage. It develops in 3 to 7 days in mid-summer.

The infective larvae, so-called because they are now ready to infect other sheep, cannot feed but they climb up and down the herbage waiting to be eaten by a grazing animal.

When they are swallowed, the infective larvae fix themselves to the lining of the stomach or intestines, and during the course of 2 to 6 weeks moult and grow into adult male and female worms through two other larval stages (*photo 2*). They are then ready to copulate and produce another crop of eggs.

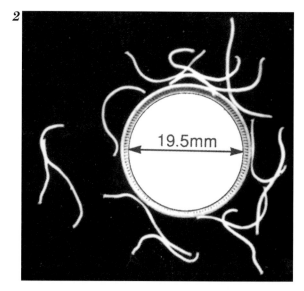

Symptoms

Lambs will tolerate a few parasites, but when large numbers are present severe symptoms result, and very often death, particularly during July, August and September.

The symptoms depend on the species of worms that predominate, the age of the sheep, and the state of its nutrition.

When large numbers of worms, e.g. the trichostrongyles, are present in the intestines, diarrhoea or scouring is the outstanding symptom (*photo 3*). The bowel becomes inflamed, the digestive products of the food are not absorbed and the affected animal becomes listless and stops thriving.

When stomach worms are present, e.g. the haemonchus (*photo 4*), the most common symptom is anaemia. The worms attach themselves to the lining of the stomach to suck blood, and cause gastritis and ulceration. Advanced cases develop a watery swelling under the jaw (bottle jaw), and the membranes of the eye become very pale, thus producing the 'white eye' condition reported by shepherds (*see* Haemonchosis – p. 71).

When, as often happens, both stomach and intestinal worms are present the symptoms are a combination of scouring, anaemia and rapid loss of condition (*photo 5, overleaf*).

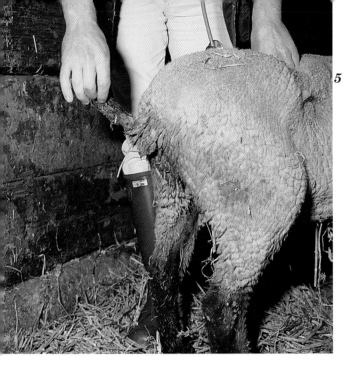

healthy sheep may pass out over 8 million eggs daily.

The basis of control, therefore, must be to avoid a build-up of infective larvae on the pastures.

The first job is to kill the adult worms in the sheep. Using one of the modern anthelmintic injections, routine dosing should be started in the spring, about a fortnight before the later lambing (*photo 6*). This will reduce the 'spring egg rise' to a minimum.

Many excellent oral anthelmintics are also available.

The lambs should be injected or dosed at 3 weeks and then at 3-week intervals until mid-June when the entire flock should be dosed again.

The dosing should be combined with sensible pasture husbandry. If the sheep are folded on small areas, for example, the hurdles should be moved at least every week,

Lambs are more seriously affected than adult sheep because the adults do possess a certain degree of immunity. As the lambs grow older, they also develop that immunity but this is never absolute and can be broken down by a gross infestation or by other stress factors. For example, parasitic disease can flare up in adult sheep in a poor nutritional state or in ewes in advanced pregnancy.

The length of survival of the infective larvae in a pasture is obviously very important. There is no general agreement but it is known for certain that an infestation will remain alive for at least 3 months, and some of the eggs and third-stage larvae (particularly those of the trichostrongyles) can survive over the winter to early spring: in fact the parasites overwinter in two ways, either as the infective larvae on the grass or as inhibited larvae in the ewe, particularly those of the haemonchus.

Prevention and control

It is obvious from the life cycle that clean sheep can only become infected by eating the infective larvae from the pasture. Maximum contamination of the pasture occurs in the spring due to the so-called 'spring rise' in the output of eggs. At that time an apparently

6

since it takes 3 to 7 days for the larvae to become infective. Where ample grazing is available, the grazing should be changed after each dosing.

Ploughing-in will not kill all the larvae, but if another crop is taken off the land before grass is resown the land should be clear.

One sensible way of cleaning up pastures is by cross grazing them with cattle or horses; the sheep larvae cannot develop in these animals and are destroyed.

So far no pasture dressing has been found that will kill parasitic larvae without destroying the herbage.

Lowland sheep

A sensible routine for lowland sheep would be:

1. Dose the ewes in April or May when they go from the lambing field to the pasture. At this time the worms which make up the spring rise will be highly susceptible to anthelmintics.

2. Dose the lambs and ewes in mid-June and move to clean grazing. In early-lambing flocks this may coincide with weaning. If this is the case, the ewes need not be dosed.

3. Dose the lambs in late August and move to clean pastures. This may be the original pasture if it has been stock-free and hayed or silaged during the summer.

Hill sheep

Between October and April the only stock are the ewes; consequently they and the returning hoggs are the reservoirs of the worm eggs.

The spring egg rise in hill sheep occurs just before lambing. Therefore hill ewes should receive their first dose of anthelmintic about a fortnight before lambing.

Because of the growth of the grass after lambing and the consequent improved nutrition of the hill ewes during the summer, and also because of their wide grazing area, they will not require dosing again until the autumn.

The hoggs returning in April from lowland wintering should be dosed before being put out to the hill and again, as gimmers, with the stock ewes, in September.

If scouring develops in the lambs at any time, they should be collected and injected or dosed immediately.

The vast majority of shepherds develop their own routine of worm dosing and are quick to spot any breakdowns. For example when occasionally young sheep develop 'black scour' during the late months an additional injection or oral dose is indicated and should be given quickly to avoid rapid loss in condition.

General hints and information

If ewes and lambs are placed on *clean* pasture there should be no need to dose the lambs before they go for slaughter. Any gimmers being kept as replacements should be dosed with the ewes in mid-July.

A *clean* pasture comprises:
● New grass after an arable crop.
● Pasture grazed by cattle the previous year.
● Grassland used only for hay and silage the previous year.

Research has shown that continued use of the same product can lead to the flock developing a resistance to that drug. It is advisable, therefore, to rotate the types of wormer using a different one each year. Your veterinary surgeon will advise you.

Anthelmintics incorporated in compound food are also available.

28 *Nematodirus* Infection

Nematodirus infection attacks young lambs suddenly and, if untreated, kills them rapidly.

1

Cause
A species of roundworm called the *Nematodirus* that can be very dangerous (*photo 1*).

It attacks in the spring – in March and April in Northern Ireland, and in May and June in the rest of Britain and especially in the Scottish border counties. Occasionally in lowland flocks the danger period extends to mid-July.

How Nematodirus differs from ordinary roundworms
The fundamental practical difference between ordinary roundworms and *Nematodirus* is that the *Nematodirus* produces a disease that differs from the typical roundworm infestation in that it is sudden in onset, hyperacute, and rapidly fatal if untreated.

The disease occurs early in the year, i.e. before the traditional roundworm trouble would normally be expected, and it affects only young lambs from 1 to 5 months of age.

Another peculiar feature is that the disease appears on pastures that, so far as the usual roundworms are concerned, would normally be considered clean and fairly safe.

What happens is shown in the diagram on p. 67. With the typical roundworm the worm eggs are laid inside the sheep and are passed out in the droppings on to the pastures. There, in the course of less than a week, they undergo three changes – into the first-stage larvae, then into the second-stage larvae, and finally into the third or infective stage.

Many of the eggs are destroyed by climatic conditions, but a considerable number reach the infective stage and, if the weather is mild, they can *keep a pasture infected for up to 6 months*. However, if there is a hot, dry summer or an exceptionally cold winter, the larvae may survive for less than 3 months.

With *Nematodirus* the eggs pass on to the pasture in the same way but thereafter the **first-, second- and third-stage larvae develop inside the eggs and this development takes approximately 2 to 3 months**. The embryonated eggs of *Nematodirus* have great resistance against sunlight, drought and frost and they remain dormant throughout the entire winter to hatch out simultaneously the following spring.

The masses of newly hatched larvae are eaten by the susceptible young lambs, and inside the lambs the larvae moult and pierce the bowel lining.

Symptoms
Sudden onset of acute diarrhoea in a group of healthy spring lambs. The affected ones

Worm development stage by stage

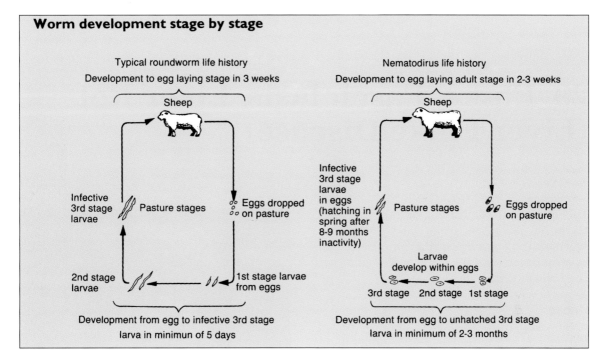

Typical roundworm life history

Development to egg laying stage in 3 weeks

Sheep

Infective 3rd stage larvae — Pasture stages — Eggs dropped on pasture

2nd stage larvae ← 1st stage larvae from eggs

Development from egg to infective 3rd stage larva in minimum of 5 days

Nematodirus life history

Development to egg laying adult stage in 2-3 weeks

Sheep

Infective 3rd stage larvae in eggs (hatching in spring after 8-9 months inactivity) — Pasture stages — Eggs dropped on pasture

Larvae develop within eggs

3rd stage   2nd stage   1st stage

Development from egg to unhatched 3rd stage larva in minimum of 2-3 months

quickly become dehydrated; they make for water, drink it to excess and stand in it. They lose weight rapidly and die in 2 to 4 days.

Treatment

The affected lambs will respond to the correct maximum dose of one of the modern anthelmintics, but the most important thing of all is to spot the condition early and get to work immediately. Any delay can cost the loss of up to half the lamb flock.

Prevention

On farms where *Nematodirus* is known to exist, it is probably best to anticipate trouble by preventative dosing at the beginning and end of the first danger month (i.e. May in most parts) with perhaps a third dose at the end of June. This triple dosing may be expensive, but nothing like as expensive as an active infection could prove.

With the continual improvement in modern anthelmintics, a single early spring dosing may be sufficient to prevent the majority of outbreaks.

Another useful preventative hint is not to use the same pasture for lambs in successive years.

Recent scientific work has shown that:

1. The *Nematodirus* egg can survive on pasture for up to 2 years.

2. Hatching of the eggs requires special stimuli in the form of a period of chill followed by a mean day/night temperature of more than 10°C (20°F).

As a result of this knowledge the Ministry of Agriculture now issues a *Nematodirus* forecast each year based on soil temperatures.

Note: A combination of *Nematodirus* infection and coccidiosis is particularly lethal.

There are two main species of *Nematodirus* – the *N. battus* (*photo 2*) seen most commonly in Britain, and the *N. spathiger* seen mainly in other countries.

2
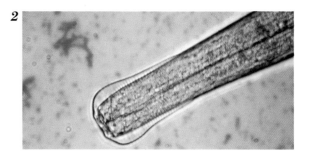

29 Fluke, Fasciolasis, Liver Rot (Liver Fluke Disease)

This is an acute or chronic disease affecting sheep of all ages (*photo 1*). It is more prevalent in some areas than in others, and usually flares up after a warm wet summer, i.e. during autumn and winter.

Cause

It is caused by the liver fluke – called *Fasciola hepatica* (*photo 2*). To understand the development and control of the disease, a knowledge of the life history of the fluke is essential.

Life cycle

For easy reference I append the life cycle in simple diagrammatic form (*see diagram A*) as well as in descriptive terms.

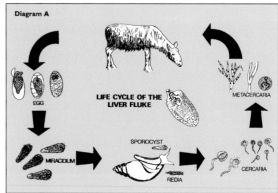

MATURE
ADULT
FLUKES
|
EGGS
|
9 DAYS
|
UP TO
5 MONTHS
|
SNAIL
(*LYMNAEA
TRUNCATULA*)
|
6 TO 7 WEEKS
|
CERCARIAE
|
UP TO
12 MONTHS

In the bile ducts of an infected sheep the mature adult flukes, which are hermaphrodite (bi-sexual), lay eggs. These are passed down the bile ducts into the intestine and out on to the pastures in the dung. On the pastures the eggs develop into the first-stage larvae – called the miracidia. This process can take place in 9 days or it can be delayed, in unsuitable conditions, up to 5 months. The miracidia swim about and are picked up by a snail called the *Lymnaea truncatula*.

Inside the snail the miracidias undergo changes into sporocysts, then into rediae and finally they develop into the second-stage larvae called the cercariae. This process takes 6 to 7 weeks.

The cercariae live only briefly: they leave snail and attach themselves to the

SHEEP
INTESTINE
|
4 TO 5 DAYS
|
SHEEP LIVER
|
PARENCHYMA
|
5 TO 6 WEEKS
|
BILE DUCTS
|
6 TO 7 WEEKS
|
MATURE
ADULT FLUKES

herbage where they quickly change intoxmetacercariae – the infective larvae. They may be eaten immediately by another sheep or they may survive for up to 12 months or longer.

When eaten by another sheep, the metacercariae penetrate the intestinal wall and in the course of 4 or 5 days, they reach the liver (*see diagram B*).

The metacercariae migrate through the substance of the liver – called the parenchyma (liver substance) – for 5 or 6 weeks, gradually increasing in size and, of course, damaging the liver in the process.

After that time, they enter the bile ducts and grow into mature adults in between 6 and 7 weeks.

In the bile ducts the adults again excrete their eggs, and the life cycle starts all over again.

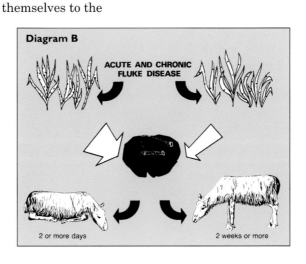

Diagram B

ACUTE AND CHRONIC
FLUKE DISEASE

2 or more days   2 weeks or more

When established in the bile ducts, adult flukes will rarely attempt to leave. Some of them have been known to survive in the same sheep for 11 years.

In the words of one of our most famous parasitologists, Dr E. L. Taylor:

> In the final host the liver fluke lives a simple life with little change. Bathed continuously in a stream of bile, at a midsummer temperature that scarcely varies a degree from one end of its life to the other, it withdraws the good nutriment that is unfailingly provided. There are no enemies to molest it; nor does it have to look for a mate to reproduce its kind, but it continues to bring forth an amazing stream of innumerable eggs in order to ensure that the species *Fasciola hepatica* shall continue on the earth.

Such is the parasite you have to destroy – and fortunately several anthelmintics will do just that.

Life cycle of *Lymnaea truncatula*

In order to understand the disease still further it is wise to study the life cycle of the *Lymnaea truncatula*.

The snails, male and female, copulate in the early spring and the females lay their eggs in March or April.

The eggs develop into adult snails in 3 months, and these in turn produce a further generation in 3 months.

Since snails live in mud and since the miracidia have to swim about to find them, it is obvious that flukes can only survive and thrive on wet land. For example, peat bogs and water springs on hill land, badly drained land anywhere, open drains, ditches and low-lying fields flanking a river which floods easily – in fact any wet or muddy land.

Symptoms
There are three types of the disease – hyperacute, acute and chronic.

In the *hyperacute* cases, sudden death is the rule and the condition is only diagnosed on post-mortem examination.

Acute fasciolasis does not begin until July and tends to get worse right up to November and December.

Several sheep, often lambs, become suddenly ill. They stand about, go off their food and are unwilling to move. When driven they fall down and refuse to rise.

The belly is swollen and painful, especially over the liver, and death takes place within a day or two (*photo 3*).

Chronic fasciolasis is seen in the latter part of the winter or early spring. It is a wasting disease which is difficult to distinguish from roundworm infestation.

Usually several sheep – adults and lambs – stop thriving. When examined, they are very thin and light and often show an oedema or dropsy underneath the jaw (*diagram C*). The

3

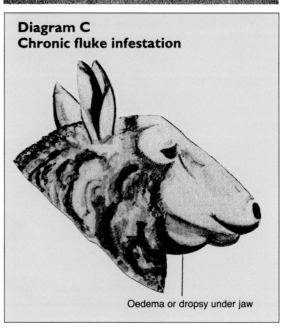

Diagram C
Chronic fluke infestation

Oedema or dropsy under jaw

mucous membranes of the mouth and eye are pale and anaemic.

Diagnosis can only be confirmed by getting your veterinary surgeon to examine the dung. Presence of the typical fluke egg or eggs will confirm fluke infestation. Withdrawing fluid from the peritoneal cavity on the spot is also a valuable aid (*photo 4*).

Treatment
There are a number of effective drugs available. Be advised by your veterinary surgeon. If you are treating in the winter or early spring when all the flukes are adults, any of the available products can be used. If treating during summer or autumn when young flukes are present, use a product that is effective against both adult and immature fluke. This will save regular repeating of treatment.

Prevention
On farms that have a fluke problem, control

4

measures as outlined in the treatment should be carried out during the winter months.

At the same time the snail population should be reduced and the best long-term way of doing this is by drainage of the danger areas.

Note: Rabbits are carriers of liver fluke so *complete* eradication is not possible unless the rabbit population is strictly controlled.

30 Tapeworms

Tapeworm infestation in lambs, although not common, is occasionally seen, and it is important to know how to deal with it. The condition is almost entirely limited to grazing lambs.

Cause
Tapeworms belonging to the family or genus called *Moniezia* (*photo 1*). The worms are wide with short segments.

The segments of the adult worm contain eggs and are passed out in the dung. The cover of the segment disintegrates exposing the eggs. The eggs (*photo 2*) are eaten by common pasture mites, and inside these mites they grow into infective larvae.

The lamb eats the mites, along with the grasses, and in the lamb's intestines the tapeworm larvae break free and develop into adult tapeworms.

Symptoms
There may be a general loss of weight and unthriftiness, but usually the segments are either seen in the dung, or a dead lamb may be found to have a large number of tape-worms in the intestine.

Treatment
Several modern broad-spectrum anthelmintics will eliminate most of the Monieza species, so routine dosing for roundworm will usually

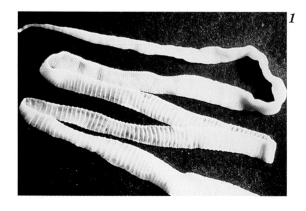

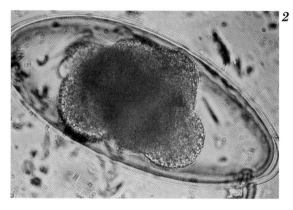

keep tapeworm at bay.

Haemonchosis

In Britain it is unusual for the *Haemonchus* stomach worm to cause disease by itself, though this occurs frequently in tropical countries. However, Haemonchosis as a clinical entity is thought to occur occasionally, affecting ewes particularly in the south of England.

The symptoms are as described on page 63, namely anaemia, bottle jaw, white eye and so on.

Obviously flockmasters in the south should be wary of the possibility of primary Haemonchosis at lambing time.

However, the *Haemonchus* in combination with the other nematodes is a major menace to all sheep populations at any time. It is very important therefore to understand the best way to control not only *Haemonchus* but all nematodes. This can be done by:

A. Clean grazing. Dose the ewes at housing or turnout and put them on clean pasture, i.e. pasture not grazed from summer the previous year by sheep or goats.

B. If no clean grazing is available in the spring there are two alternatives:

1. Dose and move around in mid-June. By then the overwintering larvae will have died, so fields used for hay or silage will be safe for both ewes and lambs.

2. Spring suppression. If lambs and ewes are treated every three weeks from turnout, with the last treatment in June, the worms should never get a chance to lay many eggs. No further treatment should then be required for the rest of the year except in heavily infected areas where continued dosing to the autumn may be advisable.

Another method of worm control is early lambing on *clean* pastures. If *Nematodirus* is not a problem and the lambs can be sent to market before the end of June, then normally there is no need to treat them before they are sold.

There are many efficient worm remedies available. Be advised by your veterinary surgeon how best to prevent the creation of resistant species of nematodes by alternating the various drugs used.

Foot Conditions

31 Foot Rot

The chief causal organism is a germ called *Fusiformis nodosus*. This microbe cannot survive outside the sheep's foot for more than 9 or 10 days, nor can it multiply in the soil. Obviously, therefore, the *only* persistent source of infection in any flock is a sheep suffering from foot rot (*photo 1*).

Two other organisms involved are *Fusiformis necrophorus* and *Spirochaeta penortha*; in fact the scientists now think the *necrophorus* and *nodosus* work together to produce the effect.

In untreated or wrongly treated cases, secondary bacteria – *Corynebacterium pyogenes, Streptococci, Staphylococci* and even *E. coli* – move in with severe and often disastrous results (*photo 2*), leading to invasion by maggots (*photo 3*).

Note. The scientists have altered slightly the names of the two main causal organisms to *Bacteroides nodosus* and *Fusobacterium nechrophorum* – though this fact is of academic importance only.

Mud or wet, badly drained pasture predispose to a disease flare up, but the mud itself will not produce foot rot unless the germs are there and they can only live on the pasture or in the mud for an absolute maximum of 21 days. It is the damage caused by the wetness, or any scratch or injury, that allows entry of the bacteria.

It is acutely contagious. If the disease is not controlled, up to 100 per cent of the flock can become infected, including the growing lambs.

Symptoms

The picture usually is one or two lame sheep followed in a few days by several more.

What usually happens in an affected foot is that a moist, mild inflammation develops between the claws (*photo 4*). This early condition is sometimes called 'scald' and is caused either by *Fusiformis necrophorus* or *Spirochaeta penortha*. Very quickly a break appears between the skin and hoof horn (*photo 5*); *Fusiformis nodosus* moves in and rapidly causes a separation of the horn from the underlying tissues. The *Fusiformis* causes death of the tissues and this leads to a characteristic stinking discharge.

Needless to say, untreated cases suffer considerable pain and lose weight rapidly, with the lameness becoming progressively worse (*photo 6*).

Treatment

As soon as lameness is detected in a flock, the problem should be tackled as quickly as possible.

Bring the entire flock, not forgetting the ram, into a concrete yard. If you haven't got a concrete yard, put them in pens floored with a thick layer of straw bedding.

Lay out one or two decent foot knives, a pair of secateurs and a sharp scalpel (your veterinary surgeon will supply the scalpel

4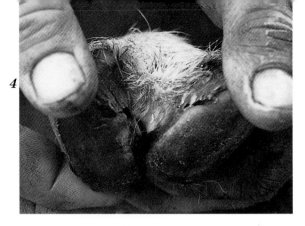

5

6

with a packet of spare blades). You also need a bucket of disinfectant for dipping the cutting tools between each foot, and a brush for cleaning the feet (*photo 7*).

Taking one sheep at a time, examine each foot using the brush, then the secateurs (*photo 8*), then the knives (*photo 9*).

Search carefully for infected animals. There are three types to look for:

1. The obvious type with lameness and under-run horn covering up the infection (*photo 10*).

2. The mis-shapen hoof in which there is a small pocket of infection under the horn,

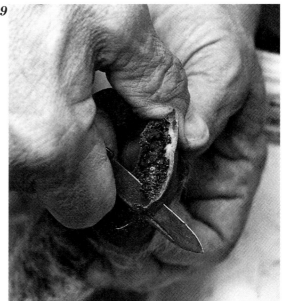

9

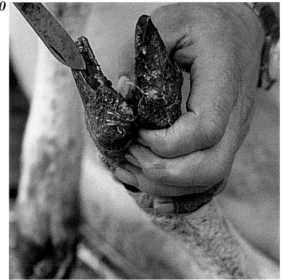

10

7

8

usually at or near the point of distortion (*photo 11, overleaf*). This is the common type of 'carrier' sheep.

3. Cases of 'scald' where the hoof is normal but the skin between the claws is hairless, mildly inflamed and moist (*photo 12, overleaf*).

After each sheep has been pared and treated, the foot parings should be swept up and the instruments rinsed in the antiseptic solution. This is often forgotten.

11

13

12

Separate all sheep in any of these three categories from the healthy stock. Now drive the clean sheep through a 10 per cent formalin, or 10 per cent copper sulphate foot-bath, keeping the feet immersed for at least 1 minute. Afterwards leave them on the dry concrete yard for an hour to two; then turn them away on to a clean pasture (i.e. a pasture which has not carried sheep for at least 21 days).

Next, set to work on the affected animals. Pare away all the dead tissue and overlying horn, paying particular attention to the front of the foot. It is absolutely vital to expose all the areas of infection so that the antiseptic or antibiotic dressing can destroy the germ.

Start with the secateurs, then the knife, but as soon as the infection is exposed, use the scalpel, always cutting from *within outwards* (*photo 13*). This simple hint will avoid bleeding and allows a much cleaner and more thorough dissection.

When all the under-run tissue is exposed, dress with one of the modern antibiotic or antiseptic dressings (*photo 14*). Most modern antiseptics will destroy the foot rot germ, but (and this is important) the old-fashioned caustic preparations such as 'butter of antimony' are inefficient and painful. They retard healing and should never be used. Ten per cent formalin is cheap and effective.

If, as sometimes happens, the foot is invaded by maggots, a special anti-fly dressing, like sheep dip, should be combined with the antiseptic or antibiotic (*photo 15*).

If the foot is extensively damaged, wrap it up in cotton wool and bandage (*photo 16*).

When all the infected sheep have been treated, turn them on to a separate clean

14

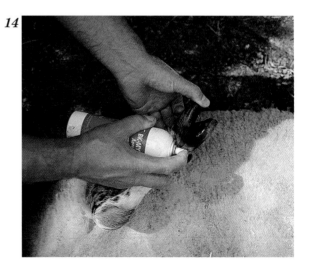

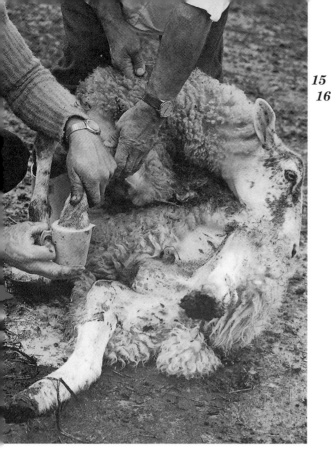

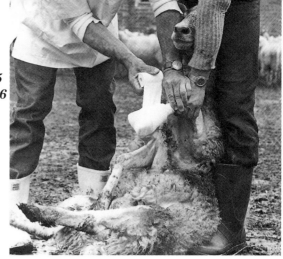

15
16

a part preventative, and as such it should be used by every conscientious sheep farmer as a vital part of commonsense routine husbandry, if only to protect the rams. The vaccines called Clovax and Footvax are similarly priced and are equally effective.

Handling

There are now several handling crates on the market which take much of the labour out of foot-paring. The one illustrated in *photos 17* and *18* is very efficient.

17

18

pasture well away from the unaffected sheep.

Examine and treat the affected animals again in 2 or 3 days, and thereafter once a week until they are cured. Two dressings will usually be sufficient.

Any new sheep, or any which leave the farm and return, should be examined, pared and segregated for at least 14 days. At the end of that time, if the feet are still healthy, they should be immersed in the formalin bath before the sheep join the main flock.

Prevention

There is a vaccine against foot rot. Two doses are given initially at an interval of 4 to 6 weeks and thereafter the immunity is boosted by a dose every 6 months.

The vaccine is excellent but not infallible, and will only prove 100 per cent effective when combined with the first-class routine husbandry described in the 'Treatment' paragraph above.

The vaccine can be used as a treatment, but obviously its main role in the future will be as

32 Additional Foot Conditions

Foot and mouth disease

During a foot and mouth outbreak, early diagnosis of the disease is usually very difficult in sheep. This is because the only obvious clinical sign may be lameness.

Cause

A filtrable virus which multiplies and spreads rapidly.

What happens

During an outbreak the sheep eat the virus (*photo 1*). It gets into the bloodstream and travels to the mouth, nostrils, feet and occasionally the udder. On these sites the virus multiplies underneath the skin to produce blisters which burst and discharge millions of the viruses on to the pasture.

The blisters on the udder, nostrils and mouth are usually small and difficult to see. In the mouth they form on the dental pad, inside the lips or on the tongue.

Symptoms

Some ewes may abort and some young lambs

1

may die. The flock becomes generally dull and many of them stop grazing. Shortly afterwards widespread lameness develops. The lameness is due to the tenderness of the ruptured blisters and to secondary ulceration and infection (*see diagram*). Close examination will reveal the blisters or ulcers around the top of, and between, the claws or at the base of the supernumeracy digits.

There may be similar lesions on the dental pad, inside the lips, or on the tongue; with a very occasional blister on the udder. Salivation may be observed.

Treatment – as a notifiable disease

Any suspect case must be reported to the

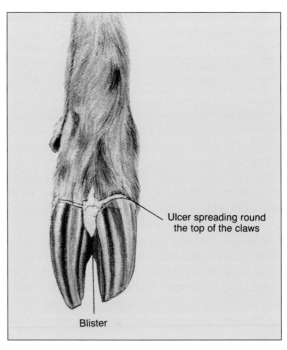

Ulcer spreading round the top of the claws

Blister

police or to your veterinary surgeon immediately. Having done so, you must on no account leave your farm until the authorities allow you to do so.

Control

If a case is confirmed, the entire flock and all contact cloven-hoofed animals will be slaughtered.

Redfoot

This is a condition thought to be limited to the Scottish Blackface breed, although it has been reported in Romney Marsh/Merino cross lambs. It affects newly born lambs *(photo 1)*.

Cause

The cause is unknown, though some scientists suspect a congenital factor.

Symptoms

The horn of the hooves becomes loose or drops off *(photo 2)* exposing the red sensitive laminae underneath – hence the name *red-foot*. Ulcers appear on the tongue and mouth lining.

The lamb is in considerable pain. It walks on its knees is unable to suck and rapidly dies of starvation.

Occasionally the eyes cloud over and ulcerate.

Treatment

The best treatment is immediate painless euthanasia (destruction).

Strawberry foot rot

This is a contagious disease affecting sheep of all ages but especially lambs.

Cause

A fungus belonging to the Dermatophilus species – the *Dermatophilus congolensis* as in mycotic dermatitis (p. 157).

Symptoms

Outbreaks usually occur when there is a wet

1

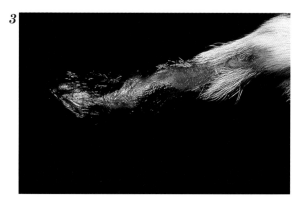

2
KNEES DAMAGED FROM WALKING ON THEM
HOOF DROPPING OFF

3

spell during the summer.

The skin between the top of the foot and the knee or the hock is inflamed and moist. It becomes thickened and scabs over. When the scabs are removed or drop off, red shallow haemorrhagic ulcers are left – hence the name 'strawberry foot rot' *(photo 3)*.

Treatment

A 3-day course of injections of mixed penicillin and streptomycin, combined with the local application of a mild caustic solution such as 0.5 per cent zinc sulphate, will clear up most cases. Long-acting tetracycline is also effective.

If large numbers are affected, they should be passed through a deep foot bath containing 0.5 per cent zinc sulphate solution, once a week for 3 weeks.

Prevention

The routine husbandry advised for the control of foot rot should be sufficient to keep the fungus under control.

Foot abscess

It is my experience that most foot abscesses in sheep are secondary to neglected foot rot (*photo 4*). However, they can and do occur independently of foot rot and it is therefore important to recognize them and treat them correctly.

Cause

The germs involved are usually the *Fusiformis necrophorus* and the *Actinomyces pyogenes*, which gain entry through minute horn cracks or wounds caused by grass seeds or awns.

Symptoms

Severe lameness. The area above the foot is usually hot, swollen and painful and the digits are spread.

Treatment

A 5-day course of penicillin injections (6cc per day of 300,000 units per cc) or 6cc of long-acting penicillin combined with local treatment such as daily soaking in hot water containing Epsom salts (one tablespoonful per 500ml/pint), or kaolin poulticing.

As soon as the abscess bursts the pain will subside, but if the course of penicillin is not given, the abscess is likely to recur.

4

BURST FOOT ABSCESS

Again, long-acting tetracycline may be used combined with, or instead of, penicillin. More and more short-acting and long-acting broad-spectrum antibiotics are becoming available, many of which are very expensive. However, if the joint is affected then the digit may have to be surgically removed.

Prevention

Control foot rot by vaccination and routine good husbandry, and foot abscesses will be a rarity in the flock.

Scald (Interdigital dermatitis)

This has already been mentioned in the chapter on foot rot (*photo 5*).

Cause

Not known, but *Fusiformis necrophorus* is usually present.

5

Treatment
Aerosol sprays of broad-spectrum antibiotic (*photo 6*) or pass the affected sheep through a footbath of 10 per cent formalin or 10 per cent copper sulphate then onto a clean pasture.

Prevention and control
As for foot rot.

6

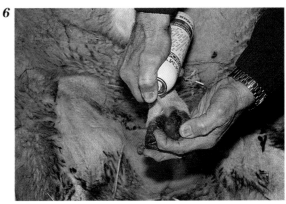

Symptoms
The skin between the claws is red, moist and painful with the pain of the inflammation occasionally severe enough to put the sheep off its legs. There may be some maceration of the skin but there is no smell and no pus. If untreated, foot rot or foot abscess may develop.

Infertility and Abortion

33 Infertility

Flushing the ewes immediately prior to mating goes a long way to preventing infertility (*photo 1*).

In Great Britain and Ireland there is no known disease that produces infertility in ewes, apart, of course, from the various abortions where retained afterbirths often lead to infertility, quite apart from the secondary effects of the abortion (*see* Abortion, p. 83). Where a number of ewes run barren on lowland farms, the fault usually lies in the management or with the tup.

The heavy feeding of ewes immediately after mating is thought to be a possible management cause.

When the tup is at fault, he is usually being asked to do too much work.

On hill sheep farms there is a known relatively high incidence of barren ewes; here the cause is generally accepted to be

nutritional – certainly no disease bacteria have been implicated.

In Australia and New Zealand ovine brucellosis produces infertility in rams.

34 Abortion

1

Abortion in ewes can be a serious problem (*photo 1*). There are a number of different types of abortion, and every flockmaster should be aware of these and should know how the various bacteria get into the flock and what damage they are likely to do. In addition, he should know exactly what precautions he can take for subsequent lambings.

First of all, there are the specific abortions, i.e. those caused by specific bacteria or organisms. For easy reference I have tabulated what I consider to be the essential practical information for the flockmaster.

Specific abortion	Cause and symptoms	How it gets into a flock and other important facts	What the shepherd should do
Enzootic abortion (Kebbing abortion)	A bacterium called *Chlamydia psittaci.* Generally no sign of systemic disease in the flock but the aborting ewes may be dull and off food for several days before aborting; and often they retain their placenta, leading to metritis.	Purchase of infected sheep of any age including lambs. Within the flock the disease is spread only at lambing time because the organisms are excreted at the time of abortion and in the vaginal discharges which clear up in three weeks. ***The infection is not spread by the ram during service.*** Enzootic abortion remains in the flock once it is introduced. The abortion rate in unvaccinated flocks will vary from 5 to 25 per cent of all the ewes, abortions occurring 2 to 3 weeks before full term.	Make use of the first-class, recently improved enzootic abortion vaccine. It produces a 3-year immunity and all the breeding ewes should be vaccinated in the late summer prior to, or at tupping. In the following seasons any replacements must also be vaccinated. During an outbreak the vaccine can be given as a preventative, though to be effective it has to be used at least 2 months before lambing. Keep any pregnant members of the family or helpers well away from a

Specific abortion	Cause and symptoms	How it gets into a flock and other important facts	What the shepherd should do
		Following abortion a strong immunity is established. This means that most ewes that abort will lamb normally in subsequent years so there should be no panic selling.	suspect flock because the Chlamydia is **very** dangerous to pregnant women: *do not let them handle the vaccine.*
Vibrionic abortion (Vibriosis)	An organism called *Campylobacter fetus* (originally known as *Vibrio foetus*).	Purchase of infected carrier sheep that excrete the bug in their faeces. Another source of infection is birds such as crows, magpies etc. The infection is taken in by the mouth and is not spread by the ram. When first introduced the disease can cause up to 60 per cent of the ewes to abort, though in most of the infections I've had to deal with the losses have been between 10 and 20 per cent, with the abortions occurring about 6 weeks to a month before term.	There is no vaccine as yet available in Britain so the ideal is to keep the flock self-contained. One attack gives a long immunity and doesn't appear to affect the ewes' fertility for the following season. However, 'carriers' will be created, and any bought-in replacements will probably become infected and may abort their first crop. Hence the advice to become self-contained if possible. A successful vaccine is used in the USA.
Salmonella abortion Once Salmonella has been identified the local authority should be notified under the Zoonosis order because of the danger to humans.	Several different strains of Salmonella bacteria including *Salmonella abortus-ovis* (seen mainly in the south of England), *S. typhimurium*, *S. dublin* and *S. montevideo*. Abortions occur mainly in the latter half of pregnancy (up to 6 weeks before term). With all the Salmonellae except	Purchase of infected but symptomless carrier sheep of any age. Occasionally the bacteria are carried by vermin and birds. As in Vibriosis, the germ is eaten and is not spread by the ram. First year losses average about 10 per cent, though severe attacks can bring the losses to 40 or 50 per cent.	Again, if at all possible, keep the flock self-contained. One attack confers a powerful immunity and the majority of ewes will breed successfully the following year. During an outbreak isolate the sick and aborting ewes, and inject selected cases with long-acting antibiotics. Subsequently spread the natural immunity by mixing the 'carrier' ewes with the non-pregnant

Specific abortion	Cause and symptoms	How it gets into a flock and other important facts	What the shepherd should do
	S. montevideo, the affected ewes are obviously ill before and at the time of abortion, with severe diarrhoea, a high fever of 106–107°F (41–41.6°C), pneumonia, or even septicaemia. Few of the aborted lambs survive, often developing fatal pneumonia or septicaemia.		breeders well in advance of the next tupping.
Toxoplasmosis abortion	A protozoan parasite called the *Toxoplasma gondii*. Symptoms vary according to when infection is picked up. If early in pregnancy the foetus may be resorbed and the ewe will return to the ram. If late in pregnancy an infected but normal and immune lamb usually results. However, infection in mid-term (60–120 days) can be disastrous. It will cause death of the foetus and abortion or mummification around 40 days later in 100 per cent of cases. But immunity following infection is strong and ewes never abort more than once.	At one time it was thought that the source was carrier sheep but it is now firmly established that the ***only source of infection is the oocysts of the* Toxoplasma gondii *that are passed out by infected cats*** – apparently the only host in which the life-cycle of the parasite can be completed. The oocysts (in the cat's faeces) contaminate the cereals, hay and bedding and are carried by vermin. (It is very important to remember that toxoplasmosis is very dangerous to pregnant women.) There is also some evidence that toxoplasmosis may be transmitted by the tup.	Keep all cats away from cereals, hay and bedding – pelleted concentrates in sacks are likely to be safer. Continually control all vermin. Once the specific cause has been established isolation is not necessary but it is always worthwhile following the advice of the veterinary surgeon in charge in case there is a mixed infection. Obviously the recovered ewes should not be sold. Prevention is the real answer and a two-season protection can be obtained by a live vaccine called Toxovax, used in consultation with the veterinary surgeon.It is very important to remember that toxoplasmosis and the vaccine against it are dangerous to pregnant women.

Specific abortion	Cause and symptoms	How it gets into a flock and other important facts	What the shepherd should do
Q fever abortion	A Rickettsial organism called *Coxiella burnetti*, not common in sheep as a clinical disease, but may act as a source of infection for man. Transmission is by direct contact.	Again, the causal agent can be spread by several carriers, e.g. vermin, ticks and man. In the comparatively rare sheep outbreaks abortion losses can be up to 10 to 15 per cent, and pneumonia and eye infections may develop as well as the abortion.	Keep the flock self-contained. A powerful immunity develops, but any incoming sheep will be susceptible to infection.

In addition to specific abortions there are the non-specific ones – those not due to microbes. Strangely enough, these far exceed the losses from specific abortions.

The causes are complex and difficult to pinpoint, but probably the most important factor in keeping their numbers to a minimum is the correct feeding of the in-lamb ewes (*photo 2*) during the last 5 or 6 weeks of pregnancy (see Pregnancy Toxaemia, p. 15).

Finally, there are the indirect abortions due to certain conditions in the mother that lead to death of the lambs in the uterus. In this country such deaths occur in louping ill, congenital goitre, tick-borne fever, border disease and pregnancy toxaemia. In other countries foot and mouth disease, Rift Valley fever, blue tongue and ovine brucellosis can all cause death of the lambs inside the ewes.

Obviously, it is wise in all cases of abortion to consult a veterinary surgeon.

2

He, with the help of a laboratory, will establish the correct diagnosis and will advise accordingly.

Good management and commonsense should prevent abortion due to the stress of routine vaccinations.

35 Urolithiasis (Inability to pass urine)

Sheep affected

Males only – housed rams and wethers or intensively fed lambs on creep feed and high concentrate intake (*photo 1*).

Cause

Excess concentration of minerals in the urine, the excess arising from the rich feed. The minerals form small calculi composed of magnesium phosphate which block the urethra or urinary passage, usually at what we call the sigmoid flexure (close to the anus) or the vermiform appendage (i.e. the 'cork-screw' at the tip of the ram's penis).

Symptoms

These may appear as early as 3 or 4 months. There is obvious discomfort, straining, kicking at the belly, twitching the tail and general restlessness. Any urine passed may be blood-stained and small crystals can usually be found on the hairs of the sheath. If untreated the urethra or bladder may rupture, probably resulting in death from uraemia.

Treatment

If seen early, injections of muscle relaxants plus a change of diet may have some success but usually the obstruction has to be removed surgically by a veterinary surgeon. From my experience this can be a very difficult job since the calculi are like fine gravel or sand. However, the condition can be cured either by amputating the vermiform appendage or by making an opening into the urethra just below the anus.

Prevention

The problem, and it can be very serious in pedigree rams, can frequently be prevented by feeding sodium chloride (common salt) at 4 per cent of the concentrate mixture, or ammonium chloride at 10g per day. Perhaps most important of all is to ensure a plentiful supply of drinking water, especially in winter should troughs freeze over. Obviously the concentrates should be changed, reduced in quantity and introduced slowly.

Since the most that can be hoped for, even after successful surgery, is that the patient can be sold for slaughter, it is obviously best to concentrate on prevention, particularly when breeding valuable pedigree rams.

Research work by Lorna Hay at the Moredun Research Institute emphasizes the necessity for ample clean fresh water and the gradual inclusion of up to 4 per cent of salt in the diet to encourage water intake. In addition her recommendations are:

- The addition of 1.5 per cent of ammonium chloride or calcium chloride to the diet. This reduces the acidity and helps to control the phosphates.
- Phosphorus should not exceed 0.6 per cent of the diet (including minerals if fed).
- Calcium in the form of carbonate, chloride or sulphate should be included in the ration to ensure a maximum calcium: phosphorus ratio of 2:1.

Obviously all this will require the co-operation of both your veterinary surgeon and your feed merchant.

The ideal forage recommended, in addition to hay, is a loose meal made succulent by the addition of diluted treacle on wet beet pulp. This succulence stimulates the flow of saliva and increases phosphorus excretion in the faeces.

Other management factors recommended are:

- the non-castration of *all* male lambs;
- encouraging artificially reared lambs at weaning time to drink both milk and water from a bucket to ensure adequate fluid intake when the milk ceases.

Obstetrics

36 How Long Should Ewes be Kept?

Some time ago I asked a farming friend of mine – one of the most successful sheep farmers I know – how long he kept his ewes. He replied, 'Until they have been dead at least three days.' This laconic retort, to my mind, sums up the commonsense approach to successful sheep farming.

Provided the udder is sound and there has been no prolapse problems at the previous lambing, it is wise to keep the ewes as long as possible and that can mean many years. The ewe shown in photo 1 was 27 years old. She had had a crop of lambs every year for 24 years and her breeding career only came to an end when she required a Caesarean section.

The advantages of keeping these old ewes are obvious. Apart from the low capital outlay, the ewes acquire a powerful resistance against all the diseases encoun-

tered on that farm. For the same reason one should always breed one's own replacements.

Bought-in sheep have little or no resistance to the resident local bugs and will take several years to settle down. This fact has been illustrated again and again during restocking.

37 Natural Lambing

It is easier to kill a ewe by lambing her prematurely or roughly than by shooting her with a humane killer. If you want to keep lambing losses to a minimum, the first basic fact to remember is that it always pays to give nature a chance.

Most ewes will lamb by themselves with no trouble whatsoever, as shown in the sequence of photographs that follows.

Early signs. If the feet and head are presented, do not interfere.

Appearance of the water bag. The lamb may or may not show.

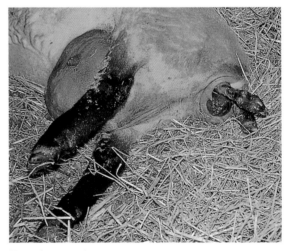

Rapid natural progress.

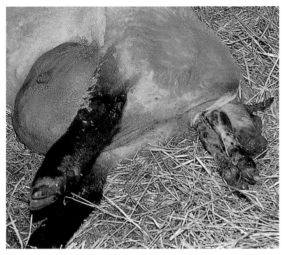

Head out and shoulders through the ewe's pelvis.

Nearly out.

Three-quarters of the way.

The final push. A healthy live lamb.

Birth of the second lamb.

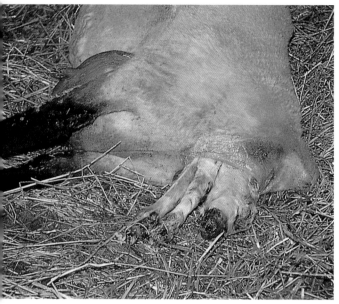

If the ewe remains prostrate, she's most likely having a second lamb. It is still wise to leave well alone.

The happy and healthy ending.

The second lamb well on the way.

It is wrong to pull a lamb away quickly even when it is straight, because the sudden change in temperature and atmospheric pressure will often kill it. Every experienced shepherd has seen this happen: an apparently normal lamb is pulled away quickly – it takes one gasp and dies. Such death is due to shock.

This is a simple but nonetheless tremendously important point well worth bearing in mind. **When the lamb is straight, never interfere unless the ewe ceases to make progress.**

38 A Guide to Good Lambing

The percentage of malpresentations in the ewe is comparatively small. In any case it is quite safe to leave a ewe in labour for at least 4 hours before examination (*photo 1*).

Apart from patience there are four golden rules at lambing time:

1. Cleanliness.
2. Getting the ewe into the correct position.
3. Lubrication.
4. Careful manipulation (*photo 2*).

Cleanliness

I know cleanliness in the field is not always easy, but I believe that, wherever possible, no examination should be attempted in the field. The suspect ewe should be loaded up and taken to the farm building (*photo 3*).

Transport is so easy on most farms that it's ridiculous to attempt interference outside. At the building, the patient should be taken out, and prepared for examination on a clean bed of straw. And, of course, it is preferable and much more comfortable if the examination can be done under a roof.

All the wool and dirt should be clipped from around the vulva (*photo 4*). This is a very useful hint, because without clipping, real cleanliness is impossible. The entire area should be thoroughly scrubbed with warm water, soap, and non-irritant antiseptic (*photo 5*).

Getting the ewe into the correct position

Lay the ewe on her side and tie the forelegs together above the fetlocks with thick bandage or string (*photo 6*). Roll her on to her back and get an assistant to stand astride her facing the tail (*photo 7*).

Holding each hind leg above the hock, the assistant should lift the hind-quarters as high as possible, and a bale or filled sack can then be put against the ewe's spine to take most of her weight (*photo 8*). In addition, a 6mm (¼in) rope can be fixed to one hind leg above the hock, passed around the back of the assistant's neck and shoulders, and tied above the other hock (*photo 9*).

I know that the hill man rarely has the luxury of an assistant, nor is it possible for him to take his patient back to a building. But in his case, an effective improvisation is easy, using the 6mm rope – with the forelegs tied, of course. This time the rope is passed over a fence post and fixed above both hocks (*photo 10*).

I think it is always worthwhile trying to get the ewe into this position because with the ewe like this, the job of examining for presentation, and of correcting malpresentation, is infinitely easier.

Lubrication

The shepherd should now again wash the vulva region, and insert copious quantities of

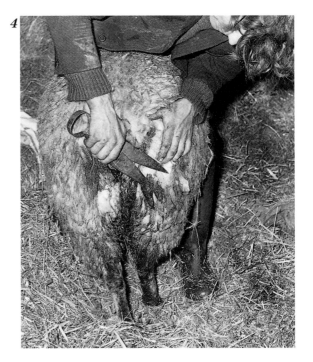

soap flakes and warm water into the vagina (*photo 11, page 96*). Some people use liquid paraffin or linseed oil; for me soap flakes comprise the ideal lubricant – and I always fill the vagina with them.

6

9

7

10

8

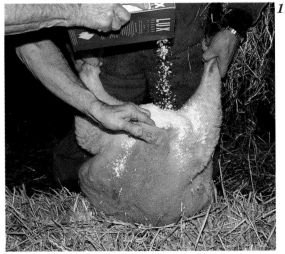

11

That done he should scrub his hand and arm, and if possible, lubricate them with an antibiotic cream and still more soap flakes (*photo 12*).

The hill shepherd, with no access to farm buildings or warm water, has to rely entirely on his lubricant cream. First of all he should fill the vagina with it and smear it over the surrounding area; then generously coat the hand and arm (*photo 13*). One important point: the old-fashioned lambing oils should be avoided because, in the main, they are irritant and do more harm than good. Every hill sheep farmer should provide himself with adequate supplies of non-irritant lubricant cream. It will be money well spent.

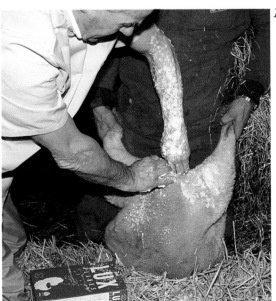

12

Careful manipulation

The hand should be introduced patiently and very slowly, taking great care not to lacerate the passage (*photo 14*). I am certain that every really experienced shepherd will agree wholeheartedly with me when I say that the ewe's inside is so delicate that one just can't be too gentle.

All the work done inside should be done with the same care, constantly renewing the lubrication when necessary. It's surprising how easily seemingly hopeless tangles can be sorted out in this position provided the shepherd is patient and gentle. *I repeat, he can't be too gentle*.

As an indication of the amount of pressure he should use, I think he should move his hand about as though inserting it between the folds of delicate silk (*photo 15*).

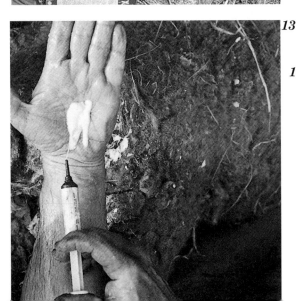

13

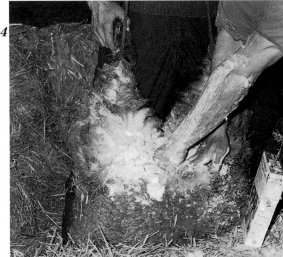

14

15 If the neck of the womb is only partially opened, the shepherd may well be dealing with the condition known as ringwomb (see chapter on Ringwomb), and he should never attempt to pull the lamb through the incompletely opened cervix. If possible, he should pass the case on to a veterinary surgeon immediately. If the presentation is correct, i.e. the two forefeet and the head of the lamb presented and the cervix fully dilated, then the shepherd should leave the ewe alone for at least another 2 hours. After that time, if the ewe still has not lambed, he should send for his veterinary surgeon.

39 Ringwomb in Ewes

Ringwomb occurs when the cervix (the entrance to the uterus) does not open properly during labour (*photo 1*). There is a true ringwomb and a false ringwomb.

In a true ringwomb, relaxation of the cervix proceeds so far and no further, despite the application of all known treatments.

In so-called false or partial ringwomb, relaxation occurs with one of the recommended procedures that follow.

Causes
Every shepherd knows about ringwomb, but not all of them understand exactly what it is, or what can or should be done when they come across it. Veterinary surgeons know what it is and what to do – but we still don't fully understand the condition because the precise cause has not yet been established. Most of us think that the majority of cases are due to some hormonal deficiency, and much

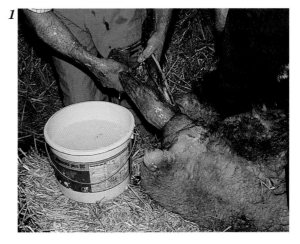

of the research work being done is directed towards proving this supposition.

Symptoms
When the ewe is examined after, say, 2 hours of ineffectual straining, only one or two

fingers can be passed through the cervix into the womb (*photo 2*), even though the two forefeet and head of the lamb may be presented quite normally. The outside edge of the cervix usually feels hard and unyielding, almost like an extended rubber ring – hence the name 'ringwomb'.

The cervix itself is made up of a group of muscles pouched together round the uterine entrance (*photo 3*). In a normal labour these cervical muscles relax and open up the entrance each time pressure is exerted from within. This pressure occurs intermittently during labour and is caused by the contractions of the muscles of the uterine wall.

Pressure on the inside of the cervix is exerted first of all by the so-called water bladder which surrounds the lamb, and thereafter, when the bladder bursts, by the pressure of the lamb's legs and head – mostly by the head – which exerts its pressure on the muscles at the top part of the cervix.

Ringwomb is not usually diagnosed until the ewe is examined, though I have found that a fairly consistent symptom is the appearance of dangling afterbirth with no sign of a lamb (*photo 4*). Such cases will probably have been

in labour for 4 or 5 hours, or even longer.

In order to examine the ewe correctly, she should be placed in the position for lambing (see p. 94) and the same precautions of cleanliness, lubrication and careful manipulation should be observed.

Is there any danger to the lambs?

In all cases of true or false ringwomb, there is some danger to the lambs, and this increases the more the labour is prolonged. The continued ineffectual uterine contractions may cause an actual separation of the placenta, or

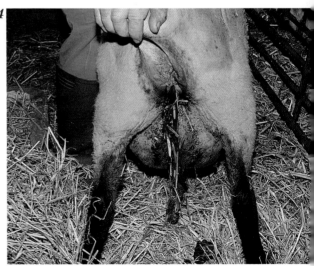

afterbirth, from the cotyledons which form the afterbirth's attachments inside the womb. When this happens, the life-giving blood supply from the mother is cut off and dead lambs are inevitable.

Treatment

If the hard, unyielding band or 'ring' is there, then the best thing to do is to turn the case over to your veterinary surgeon. I have found in such cases that the most convenient idea for both parties is for the shepherd to run the ewe down to the veterinary surgeon's hospital or premises. Even on many hill farms this idea is practicable.

If, however, as happens in some of the more remote hill farms, veterinary surgeons are not easily accessible, then there are certain things that the shepherd should try.

First of all, using plenty of soap flakes (*photo 5*) and warm water or antibiotic cream if these are not available he should pass one or two fingers through the ring and exert gentle pressure on the inside – particularly at the top – repeatedly filling the vagina and cervix with the flakes and warm water or renewing the cream and working away quietly and gently for up to half an hour or

even longer if progress is being made.

It is surprising how many 'ringed' cervices will relax and open up under this simple internal gentle digital pressure. Personally I like to keep my fingers moving round the inner ring and stopping at the top each time the ewe strains.

If *no* progress is made during the first 15 minutes of digital pressure, then an antibiotic should be injected and the ewe should be left for a further 12 hours. This, of course, presumes that the shepherd will be armed with antibiotic and syringe. In my opinion, all outland shepherds should be supplied with these by their veterinary surgeons for use in emergency. The antibiotic will help to prevent any dead lambs from putrefying.

If the sheep farm is particularly inaccessible, then the veterinary surgeon may supply a shepherd with his own more specific treatments, e.g. a muscle relaxant. This is injected intramuscularly and often acts in a spectacular manner within ½ to 1 hour (*photo 6*). In some areas certain hormone injections appear to be specific, and here again the veterinary surgeon will advise and liaise.

If patient digital pressure, time and nature, and muscle relaxants and hormones

5

6

all fail, then only two alternatives remain – emergency slaughter or Caesarean section.

Caesarean section is very successful, and most veterinary surgeons will do everything possible to keep the cost of the operation in line with the potential value of the ewe.

40 Caesarean Section

This operation is often necessary in a true ringwomb condition, and it may have to be resorted to when the foetus is oversize or abnormal or when there is an irreducible twist in the uterus. Needless to say, it should only be performed by a qualified veterinary surgeon, and I describe it here mostly for the benefit of veterinary students.

The anaesthetic
Several modern general anaesthetics can be used. Personally I prefer a local anaesthetic combined with a tranquilliser, since in my experience this technique offers the best prospect of obtaining live lambs (*photos 1 & 2*).

The site
The lower part of the left flank equidistant between the last rib and the point of the hip.

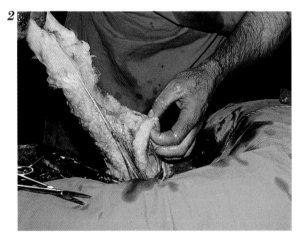

The technique
The tranquilliser is administered intravenously, and the ewe is laid on a bench on her right side. The two fore and two hind legs are tied together (*photo 3*).

The entire left flank area is clipped, shaved (*photo 4*) and thoroughly scrubbed. It is then painted with a powerful skin antiseptic and marked off by sterile operation cloths.

Approximately 20cc of a 5 per cent solution of local anaesthetic is infiltrated under the skin and into the muscles and peritoneum along the line of the proposed incision (*photo 5*). At least 5 minutes should elapse before commencing the operation. This will allow the local anaesthetic to take full effect.

Having thoroughly scrubbed up, an incision, approximately 5 or 6in (13 or 15cm) long, is made in line with the last rib and towards the lower flank (*photo 6*). The incision passes successively through the skin, the

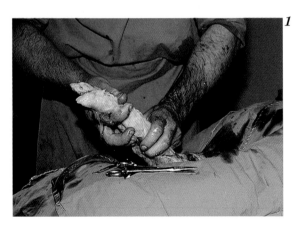

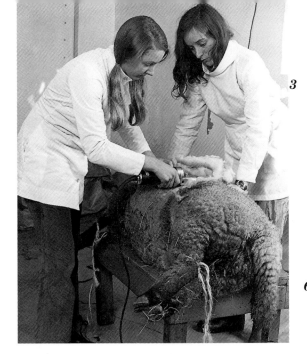

3

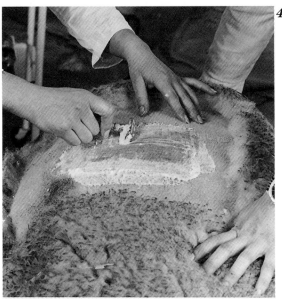

4

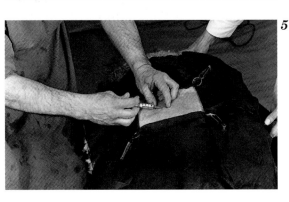

5

superficial abdominal muscle, the deep abdominal muscle and the peritoneum (*photo 7*). I find it preferable to use straight-bladed scissors, or a director, to cut through the peritoneum because of the danger of laceration of the viscera.

Both hands and forearms should be lubricated with an antibiotic oil or cream – I use intramammary antibiotics for this job.

One hand is inserted into the abdomen and underneath the nearest pregnant uterine horn (*photo 8, overleaf*). Taking great care not

6

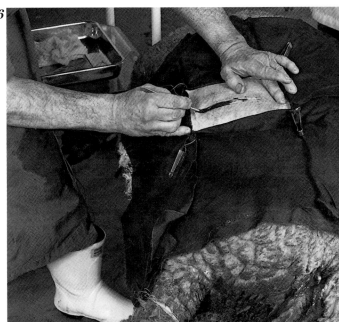

7

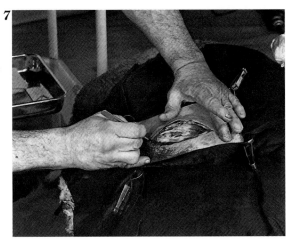

to tear the uterus, the horn is lifted up gently until it is hard against the open wound.

With the free hand a controlled incision about 4in (10 cm) long is made in the uterine wall.

Still using the free hand, and with the other still pressing the horn upwards, the lamb's hind or fore feet are withdrawn from the uterus and the lamb is lifted carefully out (*photo 9*). This lifting will bring the uterine wound outside the external incision and will leave the other hand free to assist in the withdrawal.

The remaining lamb or lambs are removed in the same way, and the lips of the uterine wound are secured with tissue forceps. If possible, the afterbirth should be removed.

The uterine wound is then closed by a single continuous Czerny-Lembert suture using double No. 3 or No. 4 catgut. I prefer this double-thick catgut because it lessens the danger of uterine tearing (*photo 10*).

The peritoneum and internal abdominal muscle are now closed by a continuous suture, again using double No. 3 or No. 4 catgut (*photo 11*). Usually I take the precaution of introducing some intramammary antibiotic into the abdominal cavity before closing the wound.

The external abdominal muscle is then brought together, again by a continuous catgut suture (*photo 12*) using double, and the skin wound is closed by two or three nylon or silk mattress sutures (*photo 13*).

A final dusting of the wound with a sulphanilamide powder and an intramuscular injection of 6cc (300,000 units per cc) of a long-acting antibiotic completes the operation (*photo 14*).

Sutures are removed in 2 or 3 weeks.

8

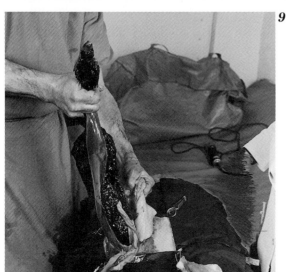

10

11

9

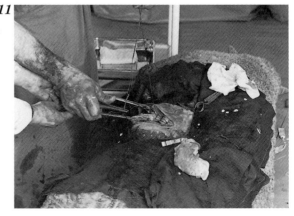

12

14

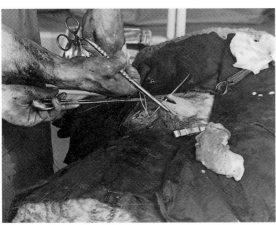

13

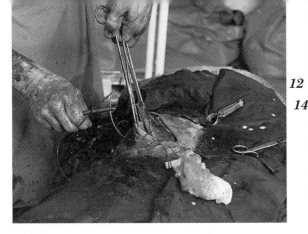

Prognosis

Excellent, especially if the lambs are alive. Even when the lambs are dead the prognosis is good, provided ample antibiotic is used in the abdomen and wound, and an intramuscular antibiotic cover is maintained for at least 10 days.

41 Prolapse of the Cervix

There are two types of prolapse – the prolapsed cervix (or neck of the uterus) and the prolapsed uterus (the entire womb). The former occurs before lambing, and the latter after. The prolapsed cervix, which is by far the more common, incorporates the ewe's vagina and occasionally the bladder also (*photo A*).

A

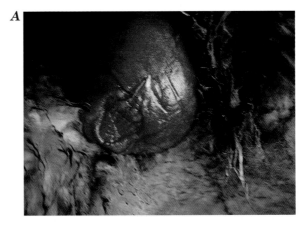

Cause

Prolapsed cervix appears to be one of the penalties of top-class farming.

In the non-pregnant ewe the rumen or first stomach alone may occupy up to three-quarters of the entire abdominal cavity, while the other three stomachs do more than their share towards filling up the remainder.

This means that, under natural conditions, there is just about enough room left for the development of a good single lamb. Twin lambs stretch the capacity to the limit, and how triplets find accommodation is little short of miraculous. In other words, the cervical and vaginal prolapse is due primarily to a lack of room (*photo 1*).

There are other contributory causes, such as excess fat, stretching of the broad ligament of the bladder, and occasionally constipation, but all these only play their part when there is overloading of the abdominal and pelvic cavities. Without that overloading, they are of no significance.

The better the management at flushing and tupping time, the greater the percentage of twins and triplets. The better the feeding, the greater the growth of the lambs. Hence my certainty that cervical prolapse in the ewe is a penalty of good farming. It is a simple matter of trying to squeeze two pints into a pint pudding basin.

Complications

When two average-sized lambs are incorporated in the prolapse, the vaginal wall

1

usually stands the strain for a reasonable time – but if the lambs are well grown and vigorous, then rupture occurs and death ensues (*photo 2*).

Treatment

The first essential is to get the ewe into the correct position for the return of the prolapse.

The technique is similar to that adopted for lambing, that is, the ewe should be turned onto her side and then rolled onto her back (the front legs can be tied together (*photo 3*)). An assistant should stand astride her facing the tail (*photo 4*) and, holding each hind leg above the hock, should lift the hindquarters as high as possible. A bale of straw or hay or a filled sack can be placed against the ewe's back to take most of the weight (*photo 5*).

If an assistant is not available, then the shepherd himself can pull the hindquarters up by fixing one end of a ½in (13mm) rope above one hock and passing the free end around the back of his neck and shoulders to fix above the other hock. He can then lift the ewe and hold it comfortably by straightening his back. It is not a bad idea to do this in any case, even when help is available (*photo 6*).

If the shepherd wants to work more freely, he can fix the rope around a post or to the top of a stone wall.

Getting the ewe into the correct position is undoubtedly the most important part of the technique of prolapse return. The weight of the abdominal contents is transferred to the diaphragm, the ewe strains with less power,

3

5

4

6

and more room is available for an easy return.

General cleanliness is very important. So use plenty of hot water, soap or, better still, soap flakes (*photo 7*) and some non-irritant antiseptic. After washing thoroughly, it is

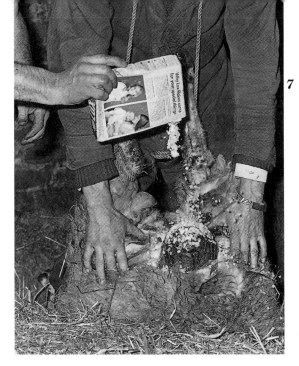

7

I sometimes use dry sulphanilamide powder as a final dressing because this penetrates the surface of the prolapse and provides ideal protection against infection (*photo 9*).

During the replacement, gentle pressure should be applied constantly and increased between strains, holding the bulk or all of the prolapse in the palm of the hand, or in the palms of both hands, depending on the size of the prolapse (*photo 10*).

And now the method of stitching. I have found the tying of the wool, the insertion of wire retainers, and the ordinary type of

advisable to dry the prolapse and smear it with a non-irritant antiseptic or antibiotic cream before replacing it in the correct position. A rough towel is ideal for drying since it removes any remaining portions of dirt without damaging the uterus (*photo 8*).

9

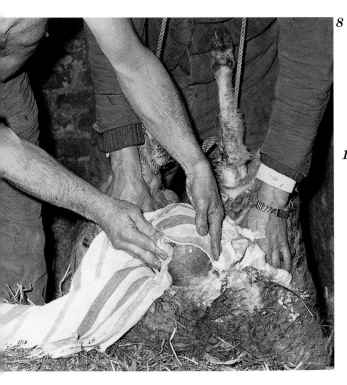

8

10

stitching completely unsatisfactory.

There is one, and only one, infallible method of stitching. It comprises a single stitch, which we call a mattress or deep retention suture (*photo 11*). A good stout needle and strong silk or nylon should be used. The method of inserting the suture is illustrated in the diagram on the right.

The needle is passed as deeply as possible through both sides of the top of the vulva (A), is brought down to the bottom of the vulva (B), and again as deeply as possible (but avoiding the urethra or hole into the bladder) is passed back right through both sides. A reef knot is now tied as tightly as possible on the side of the original entry of the needle (C).

The suture should be left in position until the ewe lambs. It is not necessary to sit up and wait for this, but it is a good idea to keep a special watch so that the stitch can be removed as soon as the lamb's feet appear.

To remove it, take the free ends of the silk just above the reef knot, pull outwards, and cut below the knot.

I have never been completely happy with the suturing technique as a precaution against subsequent prolapse for several reasons, but particularly since it involves portals of entry for infection. I am delighted therefore to report that literally as this edition goes to press I have stumbled upon what looks like the perfect answer – a

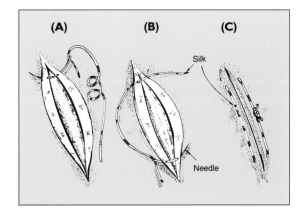

harness made of leather or webbing designed and produced by an unsung saddler (*photos 12, 13*, and *14*). This not only fixes the prolapse firmly in position, but it allows the free passage of the lamb or lambs at full term with or *without* supervision.

12

A leather harness

11

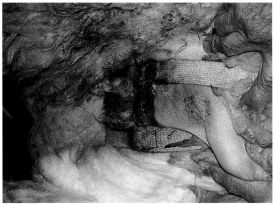

Webbing harness fitted

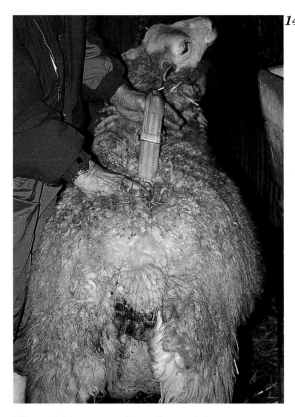

The webbing extends to the neck loop

13 Can cervical prolapse be prevented?

I don't think it is possible to prevent cervical prolapse – certainly not without the risk of losses from twin lamb disease and metabolic disorders. Flushing, ad lib hay, and an ascending plane of nutrition are a 'must' in modern, well-managed flocks.

It has been suggested (and in some parts of the country it is the practice) that hay should be restricted or withheld during the last month of pregnancy. The idea is to use as a substitute a smaller bulk feed of a high energy ration containing adequate fibre. The supporters of this theory justify it by saying that post-mortem examinations on pregnant ewes have shown that the rumen is so compressed as to be too small to accommodate hay. To this I would point out that post-mortem examinations can only be done on *dead* sheep, and the ones quoted probably died of starvation.

It is very wrong to deprive the pregnant ewe of ad lib hay. Nature will ration the intake according to available accommodation, but it is essential that sufficient natural fibre is provided to retain the concentrates long enough in the rumen, to allow the production of the fatty acids that go to keep the ewe warm and to provide energy for the growth of the lamb. Otherwise the ewe will turn to her body reserves of fat, and twin lamb disease will result.

Perhaps there is just one realistic practical step that might be tried – the avoidance, so far as possible, of a fresh new pasture during the last month of pregnancy. If there is any grass during this stage, the tendency is for the pregnant ewe to gorge.

In my opinion it is unwise to keep a ewe after she has prolapsed.

All cases should be carefully marked and culled before the next breeding season.

42 Prolapse of the Uterus

This occurs after lambing, usually within a few hours, though occasionally it can occur several days after. The entire womb, turned inside out, hangs from the vulva (*photo 1*).

Causes
There are three causes:

Calcium deficiency
When the ewe is suffering from lambing sickness (or milk fever), the deficiency of calcium causes a loss of tone in the uterine and cervical muscles. At the same time, the patient is often constipated and the reflex straining to pass the droppings brings about the prolapse (*photo 2*).

Inversion of the tip of the pregnant horn
Sometimes during labour the tip of the pregnant horn may become inverted or turned inwards (rather like the fingers of a rubber glove when it is turned inside out). When this happens, the ewe will continue to strain incessantly till the uterus prolapses (*photo 3*).

Retained afterbirth (photo 4, overleaf)
This is the usual cause when the prolapse occurs several days after lambing.

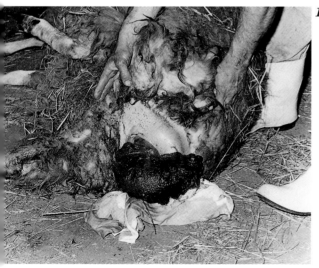
2

3

1

Symptoms

These are unmistakable. The prolapsed uterus is studded with cotyledons or 'roses', with or without the afterbirth attached. As in the cow, it can be described as a 'bloody mess' (*photo 5*).

Treatment

This is definitely a job for the veterinary surgeon.

The ewe is placed in the position for lambing and cervical prolapse return (*photo 6*). The afterbirth is removed if necessary; the entire uterus is washed very carefully with hot water containing a mild concentration of a non-irritant antiseptic (*photo 7*), and dried

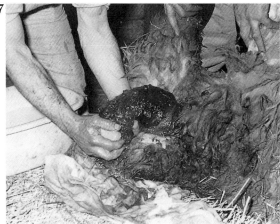

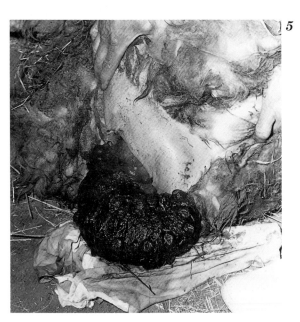

with a clean towel (*photo 8*). This is a useful hint since the rough towel removes any remaining dirt without damaging the uterus. The vulvar region is lubricated copiously with soap flakes or antibiotic cream (*photo 9*).

Dry sulphanilamide powder is massaged lightly over the entire uterine surface (*photo 10*). The hands and arms are now thoroughly washed and copiously lubricated (*photo 11*). Next the uterus is enclosed in the palms of the hands (*photo 12*) and is then very gently

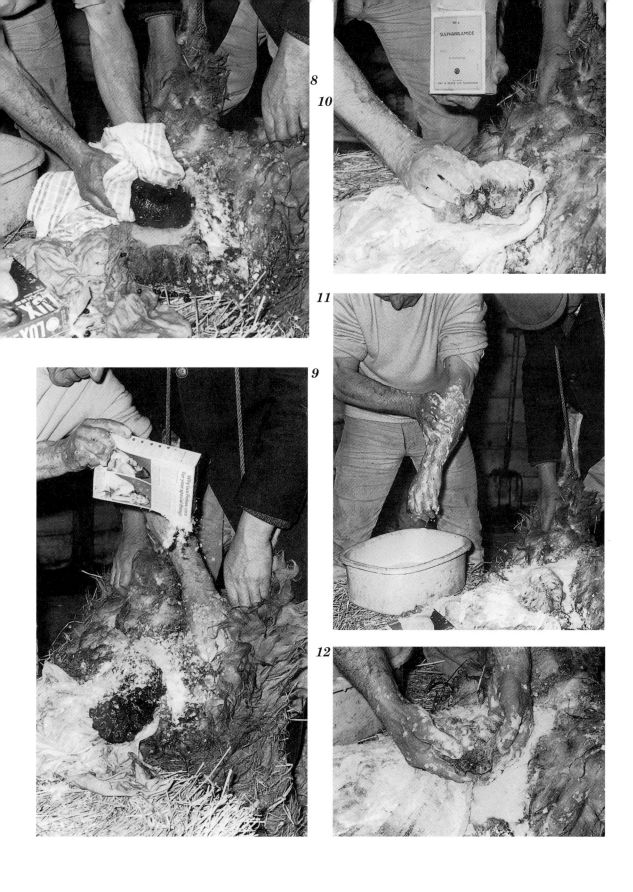

replaced between the ewe's strains.

When back in position, it is essential to make sure that the tips of the uterine horns are turned fully backwards (*photo 13*).

The vulva is then sutured in the same way as for cervical prolapse (*photo 14*).

After-treatment

If not given before the prolapse is returned, 100cc of a 20 per cent solution of calcium borogluconate should be injected subcutaneously and 6cc of penicillin or streptomycin given intramuscularly (*photo 15*).

15

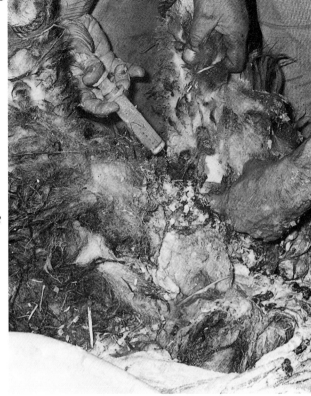

13

The antibiotic injections should be continued daily for at least 5 days, or at least a long-acting antibiotic should be given initially.

14

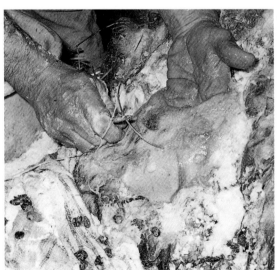

Do the ewes recover?

With modern antibiotic therapy the majority survive though, as with the cervical prolapses, it is wise to mark the ewes and cull them before the next breeding season.

The spontaneous type of prolapsed uteri and that associated with lambing sickness cannot be prevented, but that due to retained afterbirth can – by daily or long-acting injections of penicillin or streptomycin until the afterbirth is voided.

In fact all cases of retained afterbirth in ewes should be treated thus as a routine. If not, septicaemia is an even greater hazard than prolapse.

43 Dealing with Malpresentations

Caudal analgesia (above)

This makes the correction of ovine malpresentations much easier on both the operator and the ewe, but the technique requires a high degree of skill and should only be practised by a veterinary surgeon.

When the two forelegs and head of the lamb are not coming first, then the shepherd is dealing with a malpresentation (*photo 1*). If, as in most cases, the shepherd's hand is a large one, and provided professional help is readily available, it will no doubt pay handsomely to call in the veterinary surgeon.

For the remote sheep farmer, and for the shepherd with a small hand, a gentle touch and a great deal of experience, the following professional hints should be of considerable value.

Never attempt to correct a malpresentation with any part of the lamb in the passage. If possible – and this applies in the vast majority of cases – always, using copious lubrication, gently replace the impaction into the womb, where there is much more room to move about and where the danger of

1

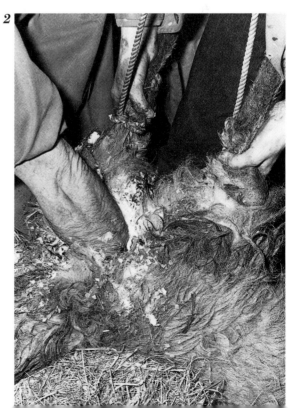

2

inflicting damage to the mother is correspondingly lessened (*photo 2*). The ewe, of course, should be placed and kept in the upended position, until the presentation is corrected.

The following sequence of photographs shows the general routine for dealing with malpresentations.

A suspect malpresentation with a dead lamb a possibility because of the dark, abnormal colour of the placental contents. After the hands and arm are thoroughly washed and well lubricated, a gentle preliminary examination may be made with the ewe in the recumbent position (i.e. lying on her side).

The first essential is to get the ewe into the correct suspended position. Wash the entire vulva area with hot water, soap and a non-irritant antiseptic, and copiously lubricate with soap flakes the hand, the arm and the ewe's vagina.

Now the careful internal examination. In this case it revealed an abnormally large head turned back and two large feet belonging to the head. The lamb was dead.

Preparing to correct the malpresentation: this will be done using two cords, and here the first cord is looped ready to be placed over the lamb's head.

Inserting the loop of the first cord over the head of the lamb to just behind the lamb's ears.

The head and both forefeet of the exceptionally large dead lamb secured ready for delivery.

Having secured the head and one foot, preparing the loop to secure the second foot.

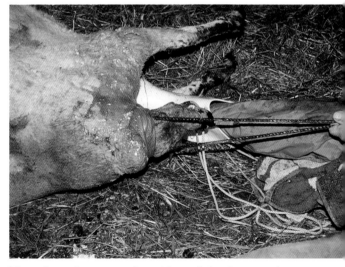

Now place the ewe on her side and, having eased both forefeet forward one at a time, ease the head gently and patiently through the vulva.

Now the forelegs of the lamb are brought further forward one at a time.

The successful removal of the oversize lamb without damaging the mother – the result of patience, lubrication and careful manipulation. The ewe made a complete recovery without complications.

Easing the lamb's shoulders, one at a time, through the vulva.

The final precaution after bad lambing – injecting the ewe with long-acting antibiotic.

44 Common Malpresentations

Head back

Both forelegs are usually in the passage, but the head is turned right back into the uterus (*photo 1*). The correct procedure to deliver the lamb is illustrated on the model and is as follows:

First of all, secure both forefeet with two lengths of strong white cord or sterile string (*photo 2*). If the cord or string hasn't been boiled, then at least it should be coated with antibiotic cream. Then very gently, and using lots of soap flakes and warm water, push the lamb – between the ewe's strains – back into the womb. Grasp the head and turn it, and, with the finger and thumb in the eye sockets, ease it

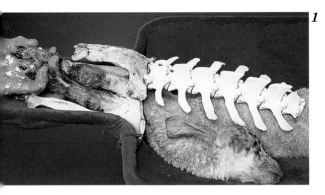

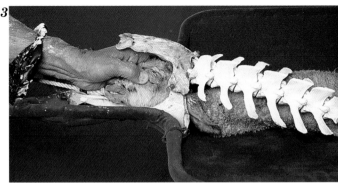

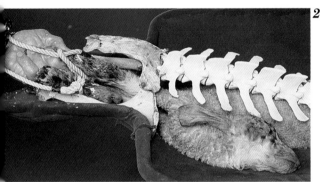

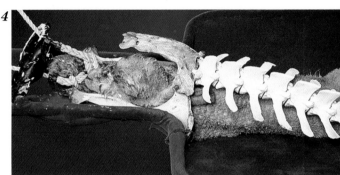

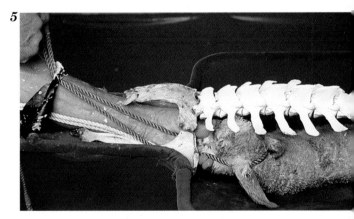

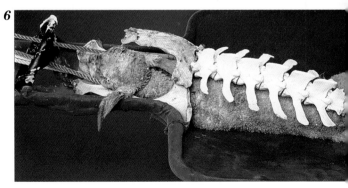

forward into the anterior vagina (*photo 3*).

Now bring the legs forward one at a time (*photo 4*), and as soon as the lamb is presented correctly, lower the ewe on to her side so that she can give the maximum of assistance with her straining. Proceed to help in the delivery, pulling as the ewe strains.

Whatever you do, don't attempt to pull the lamb's head forward in the palm of your hand; there won't be enough room for your hand and the head.

A good tip is to pass the centre part of another piece of cord (preferably of a different colour) over the top of the lamb's head to behind the ears (*photo 5*), and use this to lever the head into the anterior vagina. Never cross the ends of the cord, and never loop it round the neck. Exert the pressure only on the crest at the back of the head (*photo 6*). Pull the forelegs forward and proceed to deliver the lamb (*photo 7, overleaf*).

The basic principle for the correction of all malpresentations is the same as that just described and illustrated.

Two lambs coming together

This can provide a puzzling tangle. There may be two heads, and one foot of each lamb (*photo 8*).

My advice for correction is this: with a length of sterile cord in the hand and using abundant lubrication, slowly replace the tangle in the uterus, exerting pressure between the ewe's strains.

Select the nearer of the two heads. Pass the centre of the cord over the top of the head to just behind the ears, bringing the free ends of the cord outside the vulva (*photo 9*). Maintain the cord in position by exerting moderate pressure. It is important to keep up this moderate pressure all the while with one hand, while working with the other, otherwise the string will slip.

Renewing the lubrication, pass the free hand on to the fixed head and from there run down the side of the neck to one shoulder, keeping the fingers hard against the lamb,

then to the elbow, knee, and finally the foot. Bring the foot carefully upward and forward into the anterior vagina (*photo 10*); this can be difficult and should certainly never be done at all hurriedly. Use care, patience, lubrication and time.

Now fix this foot above the fetlock with another piece of sterile cord (*photo 11*). Manipulating the cord inside can be quite tricky, but with a little practice any shepherd with a reasonably small hand can soon become expert.

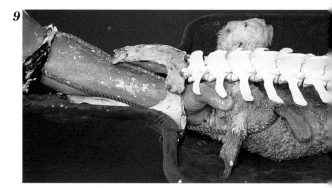

9

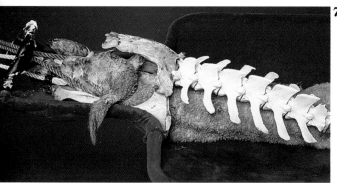

7

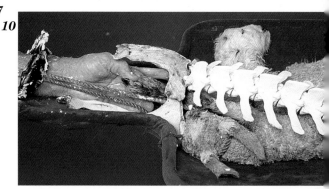

10

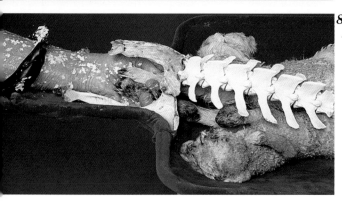

8

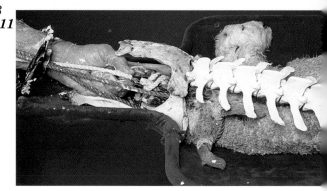

11

Once again renew the lubrication with lashings of soap flakes, and from the tied foot feel again for the knee, elbow and shoulder (keeping the fingers all the while hard against the lamb); up to the neck, to the head, and then down to the opposite shoulder, elbow, knee and foot. Bring the second foot up and forward, and secure this with another sterile cord, again above the fetlock (*photo 12*).

Next move the hand across to the other head, and push it gently back into the womb, at the same time keeping a steady pressure on all three pieces of cord (*photo 13*).

Now bring one leg at a time forward, **and having placed the ewe on her side,** patiently and slowly assist in the delivery, easing the head through the vulva and allowing the ewe to do most of the work (*photo 14*). The second lamb should present no problems.

Single head presented with one or both legs back

Follow the procedure just described for two lambs coming together, as there is often

another lamb behind (*photo 15*), and you may bring forward a wrong leg. As a matter of fact, this is the mistake that has most often been made when I'm called in.

See the sequence of photographs starting below and continuing on pages 120–21 for dealing with this presentation.

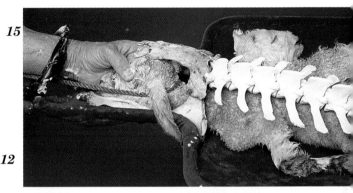

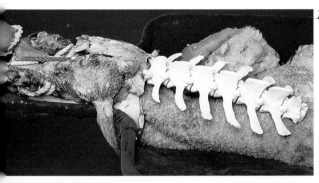

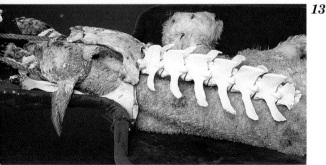

Single head presented.

Follow the prescribed procedure for two lambs coming together. Gently replace the head, bring both forelegs forward and lay the ewe on her side. When the head is in the ewe's passage the ropes can be discarded.

The lamb nearly out and alive.

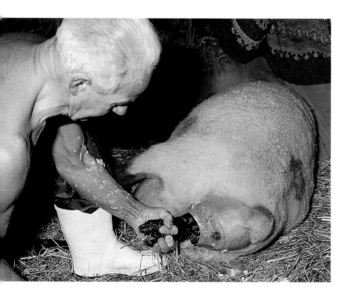

Allowing the ewe to do most of the work, exert gentle pressure. When the head appears don't panic: keep the pressure gentle and intermittent as the ewe strains.

Allow the ewe to complete the birth by herself. Clean the mucus from the lamb's mouth; the lamb would probably do this itself by vigorous head shaking.

A second lamb, fortunately presented corrected. Always examine for a second lamb and in lowland sheep for a possible triplet.

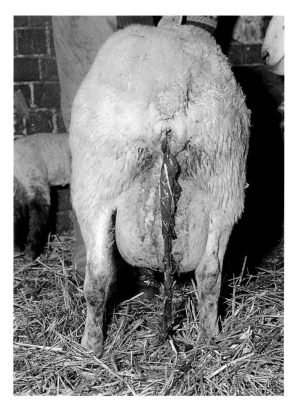

After any form of manual interference during lambing, always inject the ewe with a long-acting antibiotic. This is especially necessary when the afterbirth is retained.

Wait until the ewe is on her feet and has licked both her lambs.

Head and one leg outside the vulva

The correct procedure is first of all to place the ewe on her side and apply gentle pressure on the lamb, assisting the patient when she strains (*photo 16*). I have found that approximately 80 per cent of lambs can be delivered with one shoulder back without damaging the ewe.

If the lamb won't move with gentle pressure, however, it is wrong to pull too hard. The ewe should then be held in the position for malpresentation correction, the vulva and protruding portion of the lamb lubricated thoroughly with warm water and soap flakes, and the lamb gently replaced into the uterus, exerting pressure only between

16
18

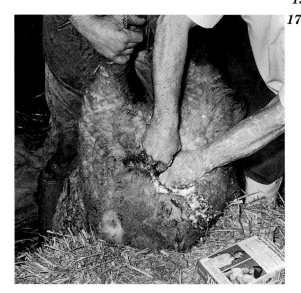

19
17

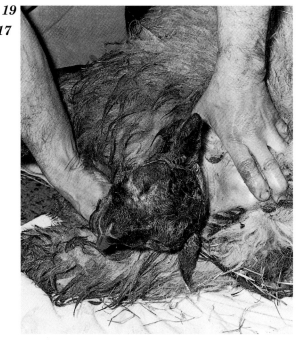

the ewe's strains (*photo 17*). Such a replacement may take some time, but with a large single lamb it is well worth the time and patience. Once back in the uterus it is usually fairly easy to bring forward the missing leg (*photo 18*).

Head out and swollen

When the head itself is out and swollen, and both legs are back, then delivery in one piece is difficult and should never be attempted unless the lamb is alive (*photo 19*).

Soap flakes are as vital as ever. The ewe should be placed in the suspended (or upended) position, and the lamb's head and inner ring of the surrounding vulva copiously lathered with warm water and soap flakes (*photo 20*). Then, with the hand or hands cupped over the head, steady pressure should be exerted, increasing the pressure between each of the ewe's strains (*photo 21*). Considerable patience is required here because the task may at first seem impossible. Gradually, however, vulva and vagina will relax and the lamb can be replaced at least sufficiently to bring one leg forward.

Naturally, of course, it is always better to bring both forelegs forward, and I have found

20

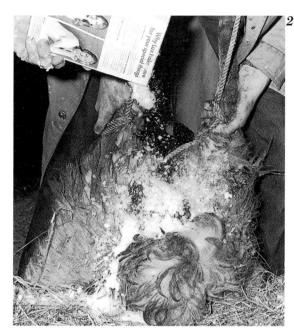

21

stump slowly, between the ewe's strains, once again using copious lubrication (*photo 25, overleaf*). When the foetus is back in the uterus, bring the forelegs forward one at a time; then replace the ewe on her side and, holding the fingers over the stump to protect the passage from laceration, deliver the lamb, with the ewe doing most of the work by herself (*photo 26, overleaf*). After the delivery, as with all bad lambing cases, inject the patient with an adequate dose of antibiotic (*photo 27, overleaf*).

Posterior presentation

In this, the hind feet of the lamb are coming first. They are identified by the feet being upside down (*photo 28, page 125*), though it is always best to examine carefully to make sure the hocks, hind-quarters and tail are there (*photo 29, page 125*), as occasionally the lamb is upside down.

The correct technique is to lay the ewe on her side and assist by pulling gently on the hindlegs as the ewe strains (*photo 30, page 125*), again allowing the ewe to do most of the work. However, with the posterior presentation, there is one very important

22

that this is nearly always possible in a ewe from the second crop onwards (*photo 22*). Large single first lambs in this position present a special problem, and these should be left to the veterinary surgeon.

If the lamb is dead and stuck with the head or head and one leg out, then the correct procedure is to cut the head off (*photos 23 and 24*), then upend the ewe and replace the

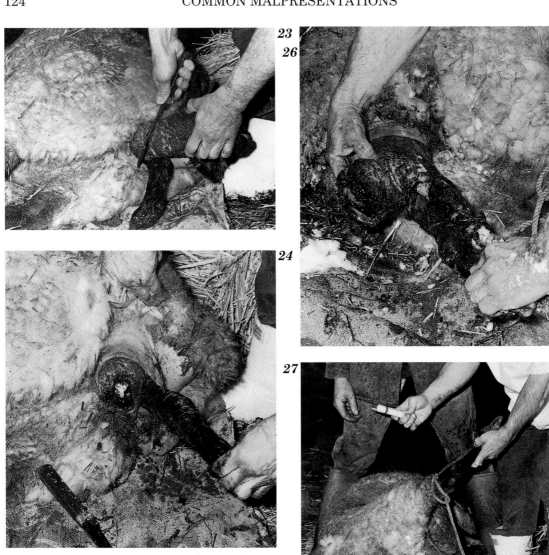

23

26

24

27

25

point to remember: as soon as the entire tail is out of the vulva, the lamb's life is in danger simply because at that stage the navel cord, which provides the vital oxygen to keep the lamb alive, is stretched to the absolute limit and is compressed between the belly of the lamb and the ewe's pelvis (*photo 31*).

The hints on how to get a live posterior lamb,

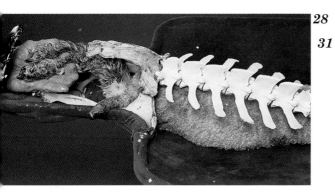

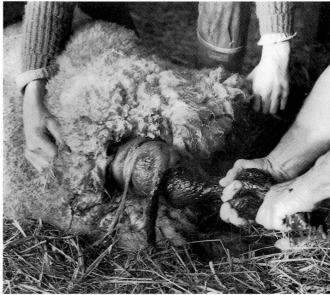

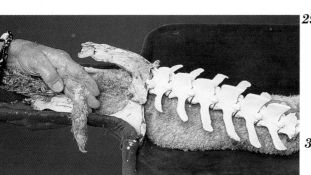

therefore, are: first of all, to rotate the lamb
through about one-quarter of a circle during the
the final withdrawal; and secondly, to make the
final pull a rapid one (*photo 32*) since any delay
will effectively cut off the navel blood supply

and will produce a dead lamb. The reason for this is that, once the navel cord breaks, the lamb has to use its own lungs to provide the oxygen, and if the head is still in the ewe when the first deep breath occurs, then the lungs are filled with mucus which suffocates the lamb. A rapid rotating delivery results in a live lamb (*photo 33*).

Breech presentation

When the hind-quarters are coming first with the tail often showing outside the vulva, the presentation is described as a 'breech' (*photo 34*).

Upend the ewe, fill the vagina with soap flakes (*photo 35*) and replace the lamb slowly between the ewe's strains (*photo 36*). Never try to force it back or the feet will rupture the womb. Now run the hand from the tail down to one hock; push the hock upwards and backwards, and the foot will come towards the vagina. Grasp the foot and bring it forward (*photo 37*).

Repeat on the other side (from tail to hock), hock upwards and backwards, and bring the second foot forward (*photo 38*). Replace the ewe on her side and assist delivery exactly as for the posterior presentation (which, of course, it now is), remembering to turn the lamb partially to ease the pressure on the navel cord and to make the final withdrawal as speedy as possible (*photo 39*).

35

36

34

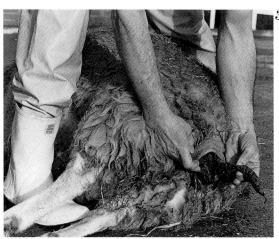

Torsion of the uterus

Just occasionally in the pregnant ewe the body of the uterus becomes twisted on its own axis (*see diagram below*). The twist or torsion may be partial or complete.

When the ewe comes to full term, she will give every sign of first-stage labour but will not get on with the job.

Examination of the vagina will reveal either a complete blockage of the passage or a corkscrew entry to the cervix or neck of the uterus.

Torsion can be easily dealt with provided it

39

37

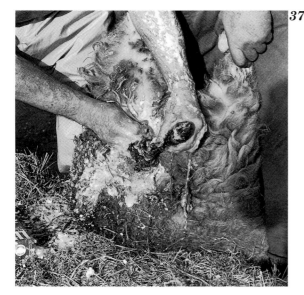

38

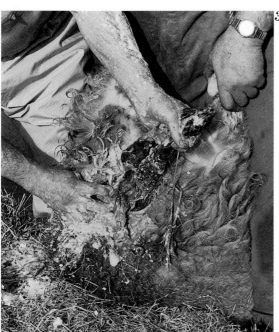

can be spotted in reasonable time. If the shepherd has missed the early labour signs, the lambs will die and putrefaction will set in.

The correct procedure is to insert your hand into the vagina as far as the twist. Then get an assistant or assistants to roll the ewe first one way and then the other (*photo 40*). When the rolling is in the correct position, the twist will be felt disentangling itself and the way to the cervix will become clear and normal.

Usually the cervix or entrance to the womb is not fully opened after the torsion correction. Muscle relaxants and antibiotics then have to be injected, and the ewe should be left for at least 12 hours or until the cervix is fully dilated.

If the lambs are alive and healthy, normal birth usually follows the torsion correction.

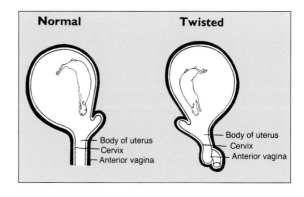

Dog-sitting position

A common presentation, and an extremely difficult one to cope with, occurs when the centre part of the lamb's spine is coming first, with the head and all forelegs pointing into the uterus (*photo 41*).

This is one for the veterinary surgeon unless the shepherd is very experienced and has a comparatively small hand.

The correct procedure is, first of all, to get the ewe into the suspended position. Fill the vagina with soap flakes, thoroughly lubricate the hand and arm, then introduce it very carefully. Keeping the fingers pressed hard against the lamb, search patiently for the lamb's tail (*photo 42*). I have found, over the years, that a dog-sitter is always easier sorted out from the hind end.

From the tail, run the hand down one hind-leg to the hock; then push the hock gently upwards and backwards and bring forward the first hind-leg (*photo 43*). Follow in again along the leg to the tail, then over to the other side to search for the other hock, always keeping the hand hard against the lamb. Bring forward the other hind-leg (*photo 44*). Finally lay the ewe on her side and deliver

41

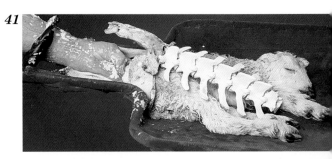

42

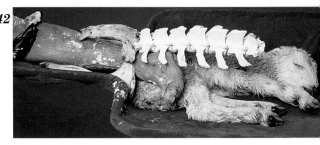

43

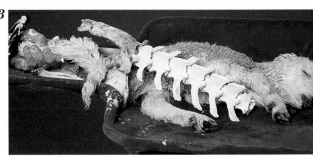

40
44

the lamb as in a straightforward posterior presentation.

To sum up obstetrics in the ewe: always give nature a fair chance and leave the ewe for a minimum of 2 hours. When examining, get the ewe into the correct upended position and observe the necessity for cleanliness, lubrication and careful manipulation.

In malpresentation, secure identifiable **45** parts and, using ample lubrication, gently replace the lamb in the womb before correcting, always moving gently and carefully and keeping the fingers hard against the selected lamb while looking for a leg or a head.

Never panic into rough groping around: patience, combined with the three golden rules of cleanliness, lubrication and careful manipulation, will resolve the vast majority of ovine obstetrical problems (*photo 45*).

45 Foetal Abnormalities

Foetal abnormalities (i.e. malformed lambs at birth) are fortunately comparatively uncommon in sheep.

Dropsical lamb

This is perhaps the commonest condition (*photo 1*). Known in some areas as 'water-belly', it has been attributed to the pregnant ewes being fed on roots. This is not the case, though obviously it is unwise to feed excessive turnips, mangolds or potatoes to pregnant sheep because of the dangers of poisoning.

Imperforate anus

1 In this condition there is no opening to the back passage (*photo 2*).

This is a congenital defect which can often be corrected by a veterinary surgeon. He

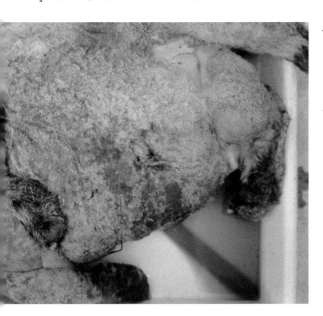

should be consulted as soon as possible.

The operation is performed under a local anaesthetic.

Entropion (Turned-in eyelids)

A condition in which the eyelids are turned in on the eyes (*photo 3*). It is seen mostly in young lambs and is undoubtedly hereditary: possibly associated with close breeding. It is being met with more and more frequently, and it can be mistaken for orf (see Chapter 66, p. 158).

I've had considerable permanent success by everting the turned-in eyelid by injecting 1cc of long-acting antibiotic into the tissue under the lid. In long-standing cases an operation may be necessary (*photo 4*).

46 Retained Afterbirth

The passage of the afterbirth in the ewe is the third stage in the natural normal birth. Fortunately the ewe's afterbirth is not often retained, but when it is there is always a reason why this should happen.

Cause

1. Premature birth (*photo 1*), the lambs coming perhaps only a few days before their full-term date, and appearing to be perfectly normal in every way.

2. Uterine fatigue after twins or triplets (*photo 2*).

3. Calcium or magnesium deficiency. Both these minerals are concerned with the contraction of muscles, and a slight deficiency of either can prevent the uterine muscle contracting sufficiently to detach the afterbirth.

4. Uterine infection, for example, vibriosis or salmonella infection, either of which can produce a high fever.

Treatment

Never attempt to remove a ewe's afterbirth manually. Manipulation of the uterus will lead to incessant straining and possibly prolapse.

The correct treatment is to maintain a systemic cover of antibiotic until the afterbirth drops. This can be done by daily injections of streptomycin or penicillin or by using a long-acting antibiotic. But your veterinary surgeon will advise you, and where necessary will no doubt supply you with the syringe and necessary drugs. Long-acting broad-spectrum antibiotics may also be used.

47 Mastitis

Mastitis simply means inflammation of the udder (*photo 1, overleaf*). Obviously, therefore, it occurs in ewes and usually after lambing or at weaning. It may be hyperacute, acute or chronic.

Cause

With the hyperacute and acute cases, over 80 per cent of them are caused by a very powerful germ called *Staphylococcus aureus* and/or the *Pasteurella haemolytica*. The chronic cases are caused by *streptococci*. The *Actinomyces pyogenes* may also be involved.

Predisposing causes

Scratches, wounds or infection on the skin of the udder and teats. The ulcers secondary to orf (see Orf, p. 158) provide an ideal field for the growth of the *staphylococci* and the *pasteurellae*.

The germ gets into the udder either via the damaged skin surface of the udder or up the teat canal (*photo 2, overleaf*).

temperature returns to normal and the ewe starts to eat again. When this happens, the gangrenous quarter sloughs or drops off during the following month or 6 weeks (*photo 3*).

In less acute cases gangrene does not develop, the affected quarter being merely hard, hot and swollen.

In chronic cases (*photo 4*) the quarter becomes 'indurated', i.e. hard. This is due to the milk secreting tissues being destroyed and replaced by hard fibrous tissues.

Treatment

The acute gangrenous case should be isolated immediately, to avoid spread of the infection, and should be injected with maximum doses of antibiotic in an attempt to save the ewe's life. Even though the affected quarter subsequently drops off, the ewe should still be able to suckle a single lamb on the remaining teat.

The less acute cases may respond

Symptoms

The hyperacute type (caused by the *staphylococci* and/or *pasteurellae* with occasional secondary *A. pyogenes*) produces the gangrenous mastitis or 'black garget'. The ewe runs a high temperature, goes off her food completely and stands with hind-legs wide apart and backwards. The affected quarter is hot, swollen and painful, but towards the teat it is ice-cold, black or blue and gangrenous. If you draw the affected teat, a watery blood-stained fluid with a peculiar sweet bread smell is emitted.

I have found that such cases may die in the first 24 hours. If they live after that, the body

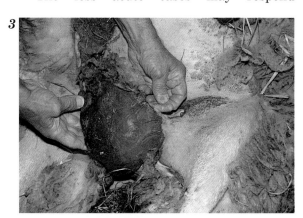

completely to a course of broad-spectrum antibiotic. An excellent example is long-acting tetracyclene.

The chronic cases are not worth treating and should be marked for culling (*photo 5*).

Control or prevention

Examine the udders of all ewes before buying them. A lumpy quarter denotes a potential carrier. Vaccines against *Staphylococcus* and *pasteurella* are available.

All clinical cases should be marked and culled.

Other sensible precautions are:

1. Clean bedding at lambing time.

2. House the ewes at lambing and for a week or two after.

3. At weaning time dip the teats and insert

a dry cow tube into each.

4. After weaning put the ewes on poor pastures well away from the sound of the lambs.

48 Care of the New-Born Lamb

Often an apparently healthy new-born lamb may appear to be dead (*photo 1*). In such a case the heart is nearly always beating, and all that is required is the stimulation or induction of the first breath.

Respiratory stimulants are now available from your veterinary surgeon; a few drops on the back of the tongue and the lamb will snatch at its first breath.

Blowing down the throat and tickling the nostril with a piece of straw, plus artificial respiration, are all valuable aids in getting the lamb going. However, I have found the most useful hint of all is to literally swinging the lamb round several times holding it by the hind-legs above the hocks, i.e. like swinging the proverbial cat (*photo 2, overleaf*). Over the years I must have saved the lives of thousands of lambs by this simple method. The greatest advantage is that it can be done immediately

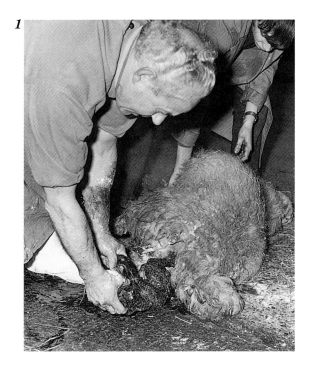

thus avoiding the loss of valuable time.

Another useful tip is to insert the thumb through the lamb's anal ring into the rectum. This will often produce that first all-important intake of air.

Reasonable warmth is an important factor in the survival of many lambs. In lowland flocks, therefore, and wherever possible in the hills, lambing should be planned in sheltered fields or in covered lambing pens (see p. 57).

If the lamb is very feeble, a teaspoonful of gin or brandy given in a 2ml syringe (*photo 3*) and a night in a warm box by the kitchen fire will work wonders.

If too weak to suck, colostrum (i.e. the mother's first milk, which not only contains antibodies against disease, but also acts as a laxative) can be given by stomach tube. It is worth while discussing the technique with your veterinary surgeon.

Once the lamb has filled his belly with his mother's milk, he rarely looks back. For this reason it is vital to check the udder of every ewe immediately after lambing, to make sure that the quarters are functioning correctly (*photo 4*). Many lambs die from starvation simply because the teats of the mother are blocked or damaged.

And, of course, never forget to dress the navel as soon after birth as possible (*photo 5*).

If lamb dysentery is a special problem on the farm, then the new-born lambs should be injected with a booster dose of concentrated lamb dysentery serum. If, despite this, diarrhoea should appear, then veterinry advice should be sought at once (see *E. coli* infection, p. 47).

If castrating and/or docking is to be carried out, the rubber rings should not be applied until the lambs are at least one day old, and are vigorous and active with several feeds of colostrum in their bellies.

49 Persuading a Ewe to Adopt an Orphan Lamb

One of the most difficult tasks in any flock is to get a ewe to take to a lamb that is not her own.

Several traditional methods are repeatedly tried, with only moderate success. During the course of my practice, I have discovered a simple and apparently infallible method, which I would like to pass on to all shepherds.

Put the orphan lamb with the ewe in a pen or loosebox, and bring the dog in (*photo 1*). The ewe's in-born maternal instinct will more or less immediately stimulate her to offer protection and refreshment to the lamb. Keep the dog in the pen for 3 or 4 minutes.

Repeat this twice daily until the ewe has settled down with the lamb. Seldom does this take longer than 4 or 5 days at the most.

The lamb adopter (*photo 2*) is a most useful alternative, and where the orphans are numerous, the method for artificial rearing shown in *photo 3* is successful.

Also of course there is the virtually infallible method illustrated in the indoor lambing system of preventing hypothermia.

50 Docking and Castration

I am convinced that both of these practices will be largely discontinued when the ever-improving nutritional standards obviate the traditional necessity.

However, since both castration and docking are still widely practised, I must admit that the best and most humane method appears to be the application of rubber rings as soon after birth as possible (*photo 1*). Ideally, the lambs should be at least 1 day old and should be vigorous and active.

Two important points to remember: where rubber rings are used, it is more than ever vital that the ewe flock should be fully vaccinated against tetanus with a booster dose immediately prior to lambing. The skin wounds produced by rubber rings can provide ideal growing grounds for the tetanus bacillus (*photo 2*).

Secondly, always make sure that both testicles are in the scrotum after the ring has been fitted (*photo 3*).

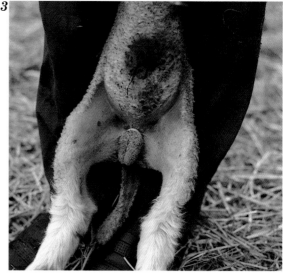

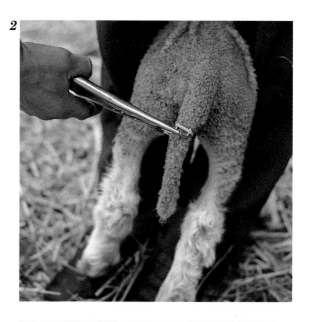

General Sheep Diseases

51 Actinobacillosis

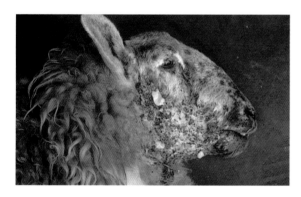

Actinobacillosis *(above)* affects the head region and shoulders of certain sheep in south Scotland. It is also seen in Sweden, Denmark and the USA.

Cause
The same germ that causes wooden tongue in cattle – the *Actinobacillus lignieresi*.

Predisposing cause
Any wound that will allow the germ to enter and grow; for example, the wounds produced by rams fighting each other.

Symptoms
One or several small abscesses about the size of a walnut, form on the lips or neck, or sometimes on the shoulder *(photo 1)*. If untreated, secondary abscesses can form, via the bloodstream, in the kidneys, lungs and liver. When this happens, the sheep rapidly loses condition.

Treatment
Using strict aseptic precautions, lance the superficial abscesses *(photo 2)* and inject the patient with antibiotic – streptomycin or penicillin – daily for 3 days. A single dose of long-acting tetracyclene is also effective.

Prevention
Since the source of infection is still not fully understood, there are at the moment no logical preventative methods.

1
2

52 Anthrax

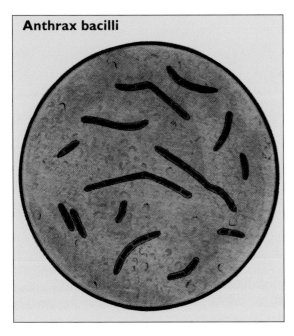

Anthrax bacilli

An acutely fatal disease of sheep occurring throughout the entire world. In Great Britain it is notifiable, and any suspect cases must be reported immediately to the police or to the Ministry of Agriculture.

Cause
It is caused by a germ called the *Bacillus anthracis*.

In this country the germ is found in imported feeding stuffs and fertilisers such as bone meal.

Occasionally the disease will flare up on a pasture in the vicinity of a previous case, or on land on which undiagnosed anthrax cases have been buried, perhaps 20 or even 30 years previously. The anthrax organism surrounds itself with a very strong capsule or spore and can survive in the soil for many years, rising to the surface on numerous occasions.

Symptoms
Death of one or several apparently healthy sheep. In suspect cases, a blood sample should be taken from an ear vein (*photo 1*) and examined microscopically, especially when a full preventative medicine plan is in operation.

Anthrax is a hyperacute disease causing a sudden high fever with the temperature rising to 108°F (42°C). The affected animal stops eating and is obviously desperately ill. It usually dies within a few hours.

Just occasionally, in the less acute form, the sheep may live for a day. If it does, it shows clear signs of acute abdominal pain and develops a bloody diarrhoea with the watery dung almost pure blood.

Treatment
If spotted early, the sheep may respond spectacularly to a massive dose of penicillin

1

(three million units) given intramuscularly (*photo 2*). Several times I have successfully treated affected sheep in flocks where deaths had occurred and where we were taking great care to check the temperatures of any of the flock that were even slightly off colour.

Post-mortem findings

It is illegal to open an anthrax carcase, but occasionally one may do so in error when suspecting hypomagnesaemia, entero-toxaemia or even lightning stroke.

The typical signs of anthrax are multiple haemorrhages throughout the body, especially in the intestines. The spleen may be grossly enlarged. If such a picture presents itself, then the case should be reported to the police or to your veterinary surgeon immediately.

Prevention

Further cases can be prevented by changing the food completely or by cutting the

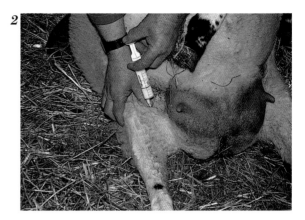
2

concentrate ration by half.

For anthrax-infected farms – that is, where the disease is known to be endemic in the pastures – a vaccine can be obtained from the Animal Health Department of the Ministry of Agriculture, Fisheries and Food through your veterinary surgeon.

53 Border Disease

This disease can occur in any flock, but it was first reported in the border counties between England and Wales, hence the name border disease. It appears to have a world-wide distribution. There also appears to be a breed disposition, for example Mules and Black-faces are more susceptible than Cheviots or Dorset Horns.

Cause

A filtrable virus called the Pesti Virus.

Source

Carrier ewes.

Mode of infection

Probably by ingestion or inhalation.

Incidence

One member, several or the majority of the flock may be affected, with the higher incidence usually in the first lamb crop. Subsequently an immunity seems to develop.

Development

Carrier ewes may show no symptoms. The virus attacks the unborn lamb, producing in many cases death of the foetus and abortion in the later stages of pregnancy.

If the lamb is carried to term and born alive, it exhibits certain characteristic symptoms.

The birth coat is hairy with kemp-like hairs rising above the wool to give a fuzzy appearance (*photo 1, right*). There is often excessive pigmentation of the skin (*photo 2*), which may persist or start to disappear after 3 or 4 weeks.

The affected lamb is unthrifty and is quickly left behind by the normal lambs in the flock. It may tremble violently and continuously except when asleep. The trembling or shaking is particularly marked in the head region and hind-quarters. If the lamb can suck, it will survive for a time but often it dies from general weakness and lowered resistance before it reaches butcher weight. The debilitated lamb suffers a sudden further loss of condition and diarrhoea before death.

The bones of the affected lambs are shorter than normal, and the skull is imperfectly developed and dome-shaped. Hence the stunted growth appearance.

Sudden death of an affected lamb is not unusual.

Treatment

There is no specific treatment. The worst affected can be culled immediately or given their chance of survival with the milder cases, but the entire lamb flock should finish up in the butchers. It is wrong to keep any for breeding.

Prevention and control

There are three sensible, practical ways to tackle the problem, bearing in mind that the

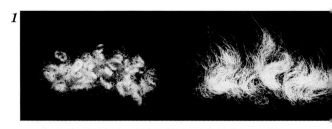

virus does not persist in the fields or farm buildings.

First of all, cull the entire flock and start again from known disease-free sources. This is by far the best method.

Secondly, continue to breed from the same ewes and rely on an immunity being built up.

Lastly, if practicable, keep the infected flock, but build up an entirely separate disease-free group, gradually disposing of the infected, one as the new flock grows.

Since it has been established that the causal agent is a virus, a satisfactory commercial vaccine will probably eventually be produced.

54 Contagious Ophthalmia (Infectious Keratoconjunctivitis)

This disease, known also as pink eye and heather blindness, can affect the eyes of sheep of all ages (*photo 1*). It is similar to New Forest disease in cattle, and is highly contagious.

Cause

It is caused by a Rickettsial organism called *Rickettsia conjunctivae* or *Mycoplasma conjunctivae*.

Where the Rickettsia comes from

The bug lurks in the eyes of many apparently normal 'carrier' sheep and becomes active only when there is some damage to the eye surface – damage caused by foreign matter, such as dust particles, long grass, chaff, irritation by flies, etc. This fact probably explains why the disease is most common in the summer and why it often flares up during dry, windy weather.

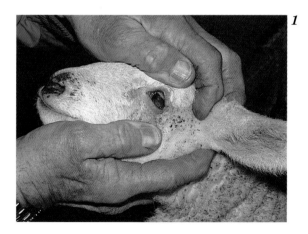

1

What happens

The damage allows the resident organism to get going, producing in the first instance a simple conjunctivitis. The Rickettsia may then pierce the surface of the centre part of the eye (called the cornea) and start to multiply.

If untreated, it forms first of all a white pin-head which rapidly increases in size. If still untreated, a yellow pointing abscess may form which eventually ruptures, leaving a filthy raw ulcer which can take up to 3 or 4 months to heal.

During this time the Rickettsia is present in the eye discharge in its most powerful form. The wind may blow such discharge over considerable distances, contaminating the surrounding pasture. This probably explains why the disease spreads rapidly and sometimes alarmingly through the flock. Of course, the strong Rickettsia is also carried from one sheep to another by flies.

A certain number of the recovered animals remain carriers and they re-infect the flock in subsequent years.

Symptoms

The first sign is a running eye. Both eyes may be affected and one or other may be closed.

If neglected, the eye will quickly film over and the sheep will be in considerable pain with consequent loss in condition (*photo 2, overleaf*).

If untreated, the eyes fill up with a yellowish pus and may ulcerate. The sheep becomes blind and tends to stay in one place,

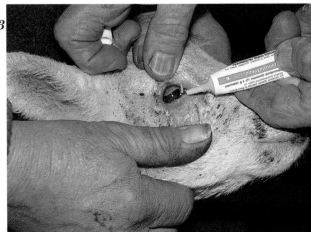

thus reducing its grazing area and contributing further to loss in condition. Permanent blindness is rare.

Treatment

Obviously, because of the rapid development and spread, it is very important to treat the disease at the earliest possible stage. There are a number of effective preparations in the form of powder, ointment, drops or emulsions, with an antibiotic as the active ingredient. A single application inserted early on will often effect a cure within a few hours (*photo 3*).

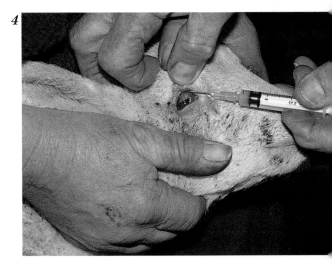

Even after the eye is 'marked' and clouded over, results can be quite spectacular, but the eye may have to be dressed daily for a considerable time. Long-acting antibiotics can be injected under the conjunctiva (*photo 4*) but this is a job for the veterinary surgeon. The scientists say that the majority of cases recover in 14 days without treatment. Personally, from practical experience, I always advise prompt attention to even mild cases.

Can it be prevented?

Unfortunately no, but it can be controlled to some extent by a constant vigil and prompt treatment. Even the young lambs must be checked (*photo 5*). I am sure that every sensible shepherd knows this and takes considerable care not to allow the spread. One obvious precaution is to isolate the affected sheep.

The reason why complete control is impossible is simply because of the persistence of

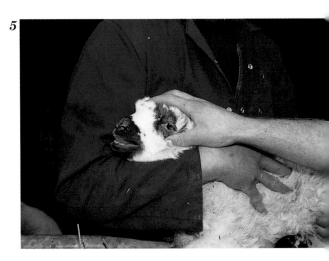

apparently normal carrier animals after an outbreak.

The only condition likely to be confused with contagious ophthalmia is xerophthalmia due to a deficiency of vitamin A (see page 31).

In xerophthalmia large batches of sheep are attacked simultaneously because the hypovitaminosis A is common to all of them.

55 External Parasites

The ending of compulsory dipping for the eradication of scab has led to a resurgence of ectoparasites on sheep, leading to a large reduction in the quality of sheep pelts. Prior to that modern dips had greatly reduced the economic importance of external parasites.

Lice – pediculosis

Lice, particularly the biting type, cause considerable damage to the sheep pelt and should be eradicated from every sheep farm. The best way to do this is by dipping as instructed by your veterinary surgeon (both pyrethroid and organophosphate dips are excellent).

Lice are an enormous problem in Australia and repeated use of *pour-on* pyrethroids has produced highly resistant strains of lice. In Britain selected dipping under veterinary supervision should avoid this.

There are three common species, two of which are illustrated in diagram A:

- The biting louse – *Damalinia ovis*, which affects the woolled parts.
- The sucking lice – *Linognathus ovillus* and *Linognathus pedalis*. These are blue-black in colour and affect mainly the lower parts of the body and legs.

Psoroptic and sarcoptic mange

Sheep scab

For some time sheep scab was virtually

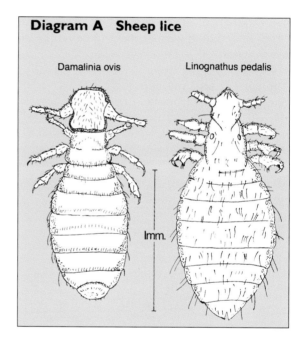

Diagram A   Sheep lice

Damalinia ovis   Linognathus pedalis

1mm.

eradicted from Britain by the compulsory routine use of highly efficient dips. Unfortunately, since the British Ministry of Agriculture has abandoned its eradication policy sheep scab has spread rapidly and widely.

It is a very serious disease, and some forms of it can cause convulsions and death within a few weeks of infestation. Because of this a farmer who takes an active case to market can be heavily fined or even prosecuted under the animal welfare legislation.

Cause and Symptoms

It is caused by two mites, the *Psoroptes ovis* (*photo 1*) and the *Sarcoptes scabei* var. *ovis*. The sarcoptic parasite, that attacks the parts least covered in wool, has not been reported in Britain for a number of years.

The *Psoroptes ovis* has, however, been persistent; it affects the woolled areas around the shoulders, back and sides, causing loss of wool and scabs or crusts surrounded by moist or wet rings (*photo 2*). It has also been found to ruin the carcase value of lambs.

Needless to say, the affected sheep are itchy and constantly rub themselves against posts and suchlike, rapidly losing weight and eventually dying, with or without convulsions if untreated.

Treatment and prevention

Sheep scab can either be controlled by correct use of an approved dip – that is, one full minute in the dip with the head under twice – or by giving two injections of ivermectin

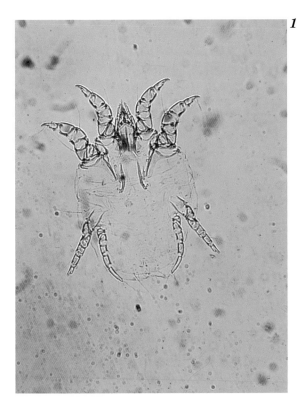

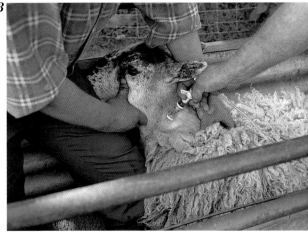

seven days apart (*photo 3*). **A single injection is not satisfactory**.

However, as inevitably progress in the formulation of new drugs continues, a later drug called Moxidectin may well supersede the ivermectin, since it has been proved to be even more effective, with a single injection lasting four weeks. Nonetheless, even with this new or subsequent drugs, two injections at an interval of ten days will most likely be recommended.

If infestation occurs during the winter or during advanced pregnancy then the ivermectin or Moxidectin injections should be used as dipping will be too stressful. Incidentally the ivermectin injections will also control keds and sucking lice but not the biting or chewing lice; though Moxidectin may well do so.

Other insecticide treatments – for example, the pour-on type – will not eradicate scab and should not be used if scab is present.

Mites can also occur in the ears of sheep without mites on the body, and it is possible that the ear mites may spread to cause scab. The ear mites cause head-shaking, and such cases should be treated by the ivermectin or Moxidectin injections once a veterinary surgeon has confirmed the diagnosis.

A simple but very important precautionary measure against reinfection is to ensure that all fencing is kept in first-class condition.

Personally I am convinced that our best hope of complete control would be the reintroduction of a compulsory eradication scheme.

Summary of new legislation to control sheep scab
As from 1 July 1997 the new British government regulations to control sheep scab are as follows.

A. Local authorities have the power to temporarily remove sheep showing clinical signs of sheep scab from common land to enable veterinary inspectors to take samples to confirm or deny sheep scab.

B. Affected and in-contact sheep must not be moved on or off premises except for treatment or immediate slaughter, or under a clearance notice issued by the local authority or a licence granted by a veterinary inspector.

C. Once a clearance certificate has been issued local authorities have the power to seize and detain sheep that have not been removed within a specified time limit. They also have the power to treat the animals at the owner's expense.

D. Sheep will be allowed back on cleared land if the owner provides evidence that the animals have been treated with an authorized produt otherwise they will not be allowed back for 3 months after the original clearance date.

Chorioptic mange

Until recently this type of mange was thought to be extinct, but it does occur occasionally.

Cause
A mite called *Chorioptes*.

Symptoms
There is some loss of hair, itchiness and the formation of crusts on the scrotum, the lower part of the legs particularly near the interdigital space, around the eye and on the brisket.

Treatment
Chorioptic mange is not notifiable and is not an important disease. Two anti-parasitic baths at 14-day intervals, or four at 7-day intervals, will clear the condition.

Demodectic mange

Although it has been recorded, demodectic mange is of no known importance in sheep.

Ticks

Ticks are common external parasites of sheep in certain areas. The important British tick is called the *Ixodes ricinus* (*photo 4*). They do not cause skin disease by themselves but are important agents for transmitting louping ill and tick-borne fever, and also, of course, tick Pyaemia.

4

Regular routine dipping plays a vital part in controlling the ticks and reducing the danger of disease transmission, particularly when allied to pasture improvement.

Blowfly myiasis (strike)

In Britain strike usually starts when the lambs start scouring either from worm infestation or too much grass. The season for blowfly myiasis is, therefore, from the beginning of June till the end of September.

Cause
The larvae or 'maggots' of the blowflies: that is, flies of the genera *lucilia*, *phormia* and *calliphora*.

Symptoms
Affected lambs lose condition rapidly and are obviously restless. Close examination reveals the blowfly maggots or larvae (*photo 5*). If neglected, septicaemia and death can result.

Treatment
Clean off all the muddy wool and as many maggots as are visible, then immerse the lamb in sheep dip; or spray the area with pyrethroid (be advised by your veterinary surgeon).

Prevention
Three dippings in June, July and September. The cause of the lambs scouring should be investigated and corrected.

Two applications of a special pour-on product are also an effective preventative.

Modern stronger dips claim greater efficiency and a more lasting effect. The dipping layout illustrated (*photo 6*) ensures immersion of the sheep for a full minute.

Commonsense, plus the use of modern dips, should adequately control *all* the external parasites of sheep.

Any veterinary surgeon or supplier will instruct on the use of dips.

56 Poisoning

In my experience poisoning is comparatively uncommon in sheep and lambs. Nonetheless, the table on the next page listing the most likely poisons and the known antidotes should be useful to sheep farmers and veterinary students.

Obviously, it is wise to be aware of these poisons (*photo 1*) and to take considerable care that the flock has no access to them.

Poison	Likely Source	Antidote
Arsenic	Weedkillers	Freshly prepared iron oxide (ferric oxide)
Carbon tetrachloride	Certain fluke drenches	Calcium borogluconate
Copper salts	Excess minerals or certain insecticides	Yellow prussiate of potassium or sodium sulphate
Lead arsenate	Insecticides	Sodium thiosulphate – intravenously
Lead	Paint – old felt impregnated with lead paint	Magnesium sulphate

57 Kale Poisoning

The random feeding of kale (*photo 1*) to sheep is dangerous. Given the opportunity, many sheep will eat to excess and poison themselves.

It is also unwise to feed excessive quantities of turnips, mangolds or potatoes to any sheep because of the danger of poisoning.

Symptoms

There is a rapid loss in condition and an acute haemoglobinuria, that is, the urine contains excess haemoglobin and is a rich red colour. Temperature rises to 104–105°F (40–40.5°C).

An acute anaemia develops. The sclera or white of the eye and the inside of the mouth become yellowish-brown in colour (*photo 2*), as also does the blood, which is thin and watery. If untreated, severe jaundice develops and death occurs in 24 to 48 hours.

Treatment

Stop feeding kale immediately. Provide ad lib

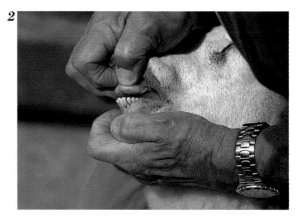

1

2

best quality hay, and inject the individual sheep with iron and vitamin B$_{12}$. Recovery can be spectacular, but deaths are the rule in many cases.

Prevention
If feeding kale, ration it carefully to a *maximum* of 1lb (500g) per head per day.

58 Other Plant Poisons

The most deadly plant poison encountered by sheep is the yew (*photo 1*). There is no antidote to yew-tree poisoning.

Contrary to the general belief, lupin and charlock are non-toxic to sheep.

Certain weeds may produce a regurgitation symptom (akin to vomiting) when the sheep are cudding. This quickly clears up when the pasture is changed or when the sheep are housed for a week or 10 days.

Bracken poisoning (bright blindness)

It is thought that bracken (*photo 2*) causes bright blindness in sheep. It does so by producing a progressive degeneration of the retina. So far the responsible factor or toxin in the bracken has not been identified. (The scientists say it occurs in areas infested by *Pteridium aquilinum*, i.e. in bracken areas.)

Symptoms
The first signs appear on the hill areas of Scotland and the north of England in September, October or November, i.e. late in the grazing season when the pasture is sparse.

The affected adult sheep are permanently blind. They hold their head questioningly forward and upwards as though constantly alert (*photo 3*).

The pupils of both eyes are circular and show little or no response to light.

1

2

Usually there is a history of accessibility to bracken.

There is no discharge from the eyes as in contagious ophthalmia, and unlike lead poisoning, gid, pregnancy toxaemia and cerebrocortical necrosis, the vacant blindness is the only apparent symptom to develop. There may be up to a 5 per cent incidence in some flocks.

3 **Treatment**
There is none.

Prevention
Burn the bracken or alternatively subsidize the diet.

Rhododendron poisoning

The classic symptom is vomiting and it is rarely fatal (*photo 4*).

59 Copper Poisoning

Sheep are the most susceptible of all the farm species to copper poisoning.

Symptoms
Acute copper poisoning, which is uncommon, can result from the sheep drinking the copper sulphate foot-bath solution. The copper damages the bowel lining producing dullness, abdominal pain, salivation and diarrhoea, and the diarrhoeic faeces are dark green in colour.

Chronic copper poisoning occurs mainly in indoor sheep during the ad lib feeding of concentrates and minerals. The affected sheep becomes dull, off its food, lethargic and may run a temperature. The membranes of the eye and mouth and the skin surface are markedly jaundiced yellow (*photo 1*).

Treatment
The administration of sodium sulphate or yellow prussiate of potassium may have some

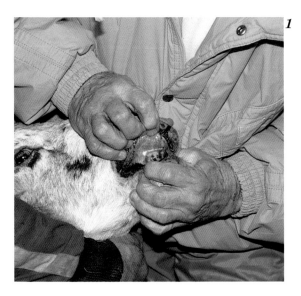

1

beneficial effect in early or mild cases but in most there is no effective treatment.

Prevention
Avoid copper in the mineral mixture; also use only established safe copper injections and oral preparations.

Your veterinary surgeon should be consulted. *I have found the 0.5 per cent copper licks safe provided there is no other source of copper in the diet.*

60 Lead Poisoning

Although comparatively uncommon, lead poisoning should be suspected in all cases of widespread illness or deaths, and especially where blindness or nervous symptoms are present (*photo 1*).

Cause
With indoor sheep, flakes of old lead paint (*photo 2*). Other possible sources are contaminated concentrates, insecticides containing lead arsenate, and old pieces of felt which are often impregnated with lead paint.

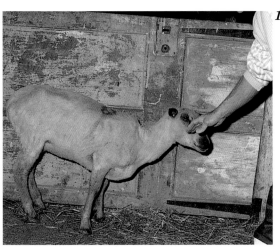
1

2

Symptoms
Are virtually identical to those described for cerebrocortical necrosis (see p. 23). The affected sheep often stands with its head in a corner (*photo 3*).

Treatment
Intravenous injections of sodium thio-sulphate (definitely a job for the veterinary surgeon) plus drenching with Epsom salts (magnesium sulphate) (*photo 4*) – two teaspoonsfuls twice daily dissolved in water and with perhaps a tablespoonful of added glucose or sugar (*photo 5*).

The magnesium sulphate is particularly useful against the lead salts (lead oxide mainly) found in paint.

Prevention
Avoid the use of insecticides anywhere near a sheep pasture, and make sure the sheep have no access to lead paint or the felt from old poultry pens (*photo 6*).

61 Sturdy or Gid (Coenurosis)

This is a brain condition, widespread throughout the United Kingdom, which may affect any sheep at any time (*photo 1*).

Cause

It is caused by the cystic stage of the tapeworm *Taenia multiceps*, which is carried by dogs and foxes. The cystic stage is called *Coenurus cerebralis*.

The sheep eats herbage contaminated with the tapeworm eggs. Inside the sheep's intestines the eggs burst to produce larvae which burrow through the intestinal wall and enter the bloodstream. The larvae travel to the brain or spinal canal, via the blood, where they grow into cysts varying in size from a hazelnut to a small orange (*photo 2*).

The cysts remain there until the sheep is slaughtered. Subsequently, if a dog eats the infected skull, portions of the cysts develop into adult tapeworms in the dog's intestines, and the cycle starts all over again.

Symptoms

The typical symptoms of gid appear when the cyst is developing in the brain, and they vary according to the site and size of the cyst.

The usual first sign is holding the head to one side (*photo 3*) and circling in that direction. The eye on the opposite side may be blind.

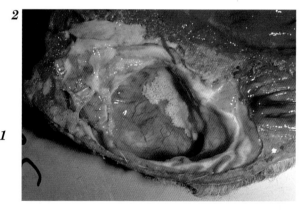

Occasionally the patient develops a high-stepping gait with the forelegs or a jerky walk.

When the cyst develops in the spinal canal, pressure on the cord produces progressive paralysis of one or both hind legs.

The patient goes off food and water and separates from the rest of the flock.

Treatment
This is a job for your veterinary surgeon. He may be able to locate the cyst (*photo 4*) and either drain it or remove it surgically.

4

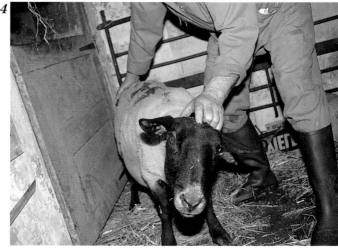

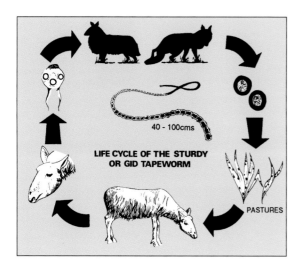

LIFE CYCLE OF THE STURDY OR GID TAPEWORM

40 - 100cms

PASTURES

Prevention
Never feed uncooked sheep's heads to dogs.

Dose all dogs on the farm against tapeworms at least twice a year using the drug recommended by your veterinary surgeon. A highly efficient injection called Dronchit is also available. During, and for 2 days after dosing, confine the dogs to their kennels and burn or bury their excreta.

As you can see from the diagram at left, the danger arises only when the tapeworm eggs are voided on the pastures. There they develop into infective larvae which contaminate the herbage.

62 Johne's Disease

This disease, sometimes called paratuberculosis, is a chronic wasting disease of adult sheep (*photo 1, overleaf*); it occurs throughout the whole of Britain. It is not common but it can affect up to 10 per cent of a closed flock.

Cause
A germ called *Mycobacterium paratuberculosis* (also known as *Mycobacterium johnei*.)

How it is spread
The bug gets into the digestive tract in food or water contaminated by dung from affected sheep. It may also be passed on congenitally.

Mycobacterium paratuberculosis produces a thickening and occasional corrugation of

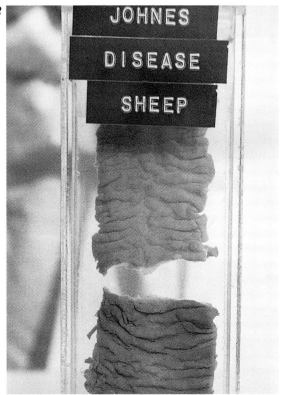

the wall of the small intestine *(photo 2)*. This takes a long time; consequently a sheep is usually at least 2 years old before it begins to show symptoms of the disease.

Symptoms
The usual picture is that one or two of the flock fail to thrive despite a good appetite. They gradually become progressively emaciated, and develop anaemia and a swelling under the jaw. The dung loses its typical pellet form, though diarrhoea is rare *(photo 3)*.

Since similar symptoms can be produced by fluke and worm infestation or by pine, the differential diagnosis should be made by a veterinary surgeon.

Treatment
The disease is incurable, so all suspect cases should be slaughtered immediately to prevent further spread of infection.

Post-mortem picture
The lining of the small intestine, particularly near its junction with the large bowel (i.e. near the ileocaecal valve) is thickened, sometimes corrugated *(photo 4)* and golden yellow in colour. This is because the Johne's germs contain a yellow pigment.

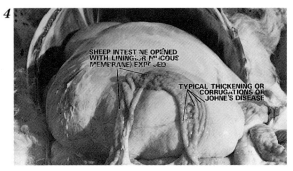

Prevention

Get your veterinary surgeon to inspect all suspect cases, and slaughter as soon as the disease is diagnosed.

All breeding ewes can be vaccinated with 0.75ml of Ministry cattle vaccine.

Raise the plane of nutrition of the rest of the flock (*photo 5*). Like most other diseases, Johne's thrives in the undernourished.

Any potential reservoir of infection such as stagnant water or ponds should be fenced off, so long as a fresh water supply can be provided.

63 Lightning Stroke

In all cases of sudden death, the possibility of lightning stroke should not be overlooked. Most farmers are insured against lightning, and a prompt diagnosis by your veterinary surgeon could ensure that you receive adequate compensation.

Symptoms

Sudden death. The carcase may lie alongside a wire fence or underneath a tree (*photo 1*), but not necessarily so. In fact by far the greatest number of animals killed by lightning stroke are found in the open field.

Post-mortem findings

A positive diagnosis can only be made by a post-mortem examination carried out by a veterinary surgeon.

The typical signs are, first of all, single marks on the surface of the skin, on the shoulder (*photo 2*) and back, or on the inside of the legs. The corresponding areas of the hide under the surface are clearly marked by extensive bruising and haemorrhages.

The marks and haemorrhages continue into the tissues under the skin and deep into

the muscles, with more acute damage occasionally evident right through the chest and abdominal cavity.

In the chest there is invariably acute congestion and also minute haemorrhages throughout both lungs, and often long strings of clotted blood in the windpipe and bronchi. The heart is contracted, with the main pumping cavities – the ventricles – nearly always empty.

The blood is not 'fluid' but it does not clot properly, the clots being soft and not clearly formed even when left for a day or two.

Sometimes a wad of grass or partially chewed hay is in the mouth, but the rest of the digestive system is absolutely normal with no signs of bloat. In fact I've never seen a bloated rumen in a true case of lightning stroke.

Obviously, it is wise to have every sudden death in the flock investigated by a veterinary surgeon.

64 Listeriosis

This condition (*photo 1*), known as 'circling disease', was at one time uncommon in Britain but it has increased dramatically during the last twenty years due to, or certainly closely associated with, the feeding of silage.

First described in New Zealand, it is common on the continent and widespread in the USA.

Cause
It is caused by an organism known as *Listeria monocytogenes*, which is thought to enter the sheep via wounds on the lips, gums and tooth base. It can also cause abortion and meningitis in humans. Apparently the germs multiply and grow in the conditions produced during the making of silage. 'Carrier' rabbits and rodents may transmit the germ, and carrier sheep excrete the bacterium in both milk and faeces.

The listeria can survive in soil, faeces and food (especially poor silage) for many months.

Symptoms
There are three types of the disease:

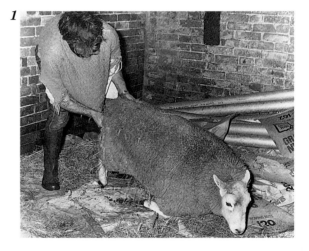

The nervous type
The affected sheep, usually an adult, runs a fairly high temperature and goes off its feed. It becomes weak on its legs and starts to walk in circles pushing against obstacles as though unable to see. The legs gradually become paralysed (*photo 2*). Often one side of the face is paralysed, causing drooling and the retention of partially chewed food in the mouth.

2 In the abortion form no nervous signs appear.

The septicaemic type
Seen in young lambs. Most of these affected are found dead, though they may survive a high fever for 24 hours.

Treatment
American reports suggest that chloramphenicol, penicillin, streptomycin and tetracyclene can produce apparent cures, but treatment must be started very early.

Prevention
If possible avoid silage feeding; if not, never feed from the sides of the clamp and discard any soil-contaminated or mouldy silage that is darker and has a much more unpleasant acid smell.

Feed the silage from troughs and keep them clean.

The abortion type
Abortion occurs without apparent general symptoms at any time after 3 months of pregnancy. Up to 15 or 16 per cent of lambs can be lost, and the condition can only be diagnosed by laboratory examination of dead lambs' blood and stomach contents.

65 Mycotic Dermatitis (Lumpy Wool Disease, Wool Rot)

Apart from strawberry foot rot already described, there is one other common fungal condition which affects the skin and wool of sheep of all ages: that is lumpy wool disease or wool rot (*photo 1, overleaf*). It occurs more or less everywhere but is probably most common in heavy rainfall areas.

Cause
A fungus or bacterium that is called *Dermatophilus congolensis*.

Symptoms
Since there are no general disease symptoms and practically no itching or baring of the fleece, lumpy wool is not usually diagnosed until shearing time. When it is present, the wool is downgraded.

What usually happens is that patches of skin become red and moist. The acute redness subsides, and when the discharge dries, yellow or black crusts are formed which are carried up with the wool producing lumpy areas (*photo 2*) – hence the name 'lumpy wool'. In adults the back, sides and neck are affected.

I have seen the condition in lambs where it seemed to attack the regions round the ears, face and neck.

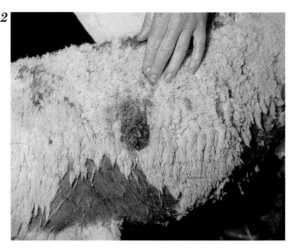

1 2

Treatment

Treatment is by dipping in a 1 per cent aluminium sulphate solution or other antimycotic drugs. The dipping should be repeated in 2 months, though with lambs a single dip is usually sufficient.

Prevention

In flocks where lumpy wool is known to occur, routine dipping under modern veterinary supervision will be well worth while.

66 Orf and Similar Conditions

Two other names for orf are contagious pustular dermatitis and malignant aphtha. It is a contagious skin disease occurring in sheep and goats, and sometimes in humans. Young sheep are mostly affected (*photo 1*).

Cause

It is caused by a filtrable pox virus – a very minute germ which can pass through a bacteria-retaining filter. The virus lies dormant in, and is carried by, many apparently healthy sheep. It lives in the tonsils and in other lymphatic glands, and in scars where it can survive for years. It can also persist in lambing areas for some time.

Where and when it occurs

It appears in all parts of the British Isles. In some areas, for example the border counties

1

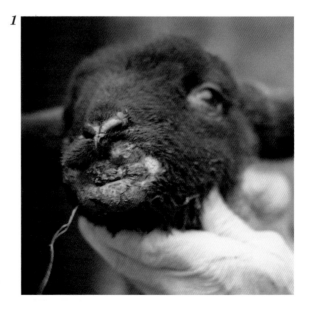

of England and Scotland, orf is persistent and a constant problem; in other parts the disease occurs sporadically, breaking out for no apparent reason.

Outbreaks can flare up at any time during the year but are most commonly seen in the spring and summer. Sheep of any age may be attacked, but for the most part orf is more prevalent in lambs up to 1 year old, probably because the older sheep acquire an immunity.

During an attack the virus gets into the bloodstream and travels to underneath the skin of various parts of the body. There it multiplies to form blisters, and these burst to form ulcers (*photo 2*) which become infected by secondary germs to produce pustules.

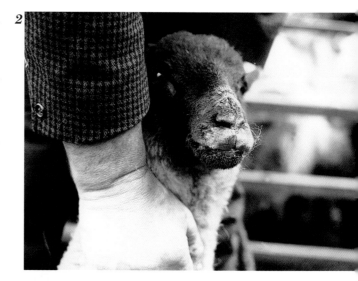

Symptoms
There are two types of orf – the benign type and the malignant type.

The *benign type,* the commonest, is characterized by the fact that most of the ulcers and pustules form round the mouth and nostrils, especially near the commissures or junction of the lips. Occasionally, similar outbreaks occur around the coronet or tops of the feet and on the teats and udder (*photo 3*), with the odd ulcer and pustule appearing elsewhere on the body.

Apart from being evil-looking and smelly, benign orf lesions can cause a loss in condition probably because the affected sheep has difficulty in eating. Occasionally secondary sepsis from the pustules may produce a fever and complete loss of appetite. But for the most part, the benign lesions are localized, and more unsightly than dangerous.

The *malignant type* is a different proposition. Almost invariably it attacks sucking lambs (*photo 4, overleaf*). Here the virus invades the entire cheek cavity, producing blisters and ulcers on the tongue, cheeks, palate, gums, inside of the lips, and sometimes throughout the whole body, with similar lesions appearing simultaneously around the top of the feet, the outsides of the mouth and lips, the genitalia and the skin on

the body surface.

During such a malignant outbreak in the sucking lambs, severe blisters and ulcers often develop on the teats of the ewes and these become infected, producing an acute mastitis which further complicates the whole miserable picture.

The cause of death in the sucking lambs is usually the infection by secondary bacteria, which is accelerated by the lowered resistance produced by the lamb's inability to suck or eat.

In both the benign and malignant types, profuse salivation may occur.

4 5

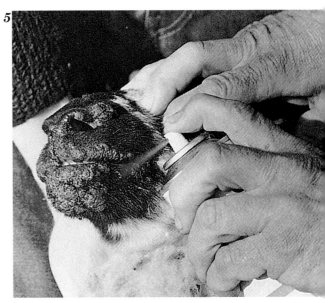

Treatment

Since there is always a possibility that the virus may become strong enough to produce the malignant type, all cases of orf – no matter how mild – must be controlled and certainly deserve careful routine attention.

It is best to isolate the affected sheep immediately and to dress the lesions daily with an antiseptic, sulpha or antibiotic cream (your veterinary surgeon will be only too happy to supply the requisite preparation).

Aerosols are particularly handy for application (*photo 5*), and I have found those containing tetracyclene most effective.

Malignant orf can only be treated under the direct supervision of your veterinary surgeon.

Prevention

Two extremely useful and inexpensive vaccines against orf are available; and in areas where the maligant type has been encountered this vaccine should be used as a routine. Elsewhere, I would say vaccination should be resorted to after consultation with your veterinary surgeon, since it is unwise to use the vaccine on clean farms.

The vaccine should be used 14 to 21 days before the outbreak is expected to occur (i.e. in March or April). This gives the antibodies against the virus plenty of time to establish themselves in the bloodstream. The latest vaccine can be used on day-old lambs and all sheep on the farm should be vaccinated.

The immunity given by the vaccine lasts for 6 to 12 months, and although it is not a complete immunity, it does prevent the development of the severe widespread lesions. In other words, it controls malignant orf and markedly reduces the lesions in the benign form.

The technique of vaccination is extremely simple. Two short superficial scratches each about 25mm (1in) long are made in the form of a cross with the vaccine-impregnated scarifier supplied with the vaccine. The site? A cleansed area of the hairless region inside the thigh (*photo 6*). As with all vaccines, the instructions supplied should always be followed carefully.

Since the orf virus can affect humans, care must be taken in handling affected sheep. In

6

man it causes hard, painful, rounded and inflamed areas which persist for up to 6 weeks.

Entropion (turned-in eyelids)

A condition which may be mistaken for orf. *See* page 130.

Periorbital eczema (Staphylococcal dermatitis)

This is another condition that can be confused with orf. It is more of an infective dermatitis than an eczema.

Cause
The chief predisposing cause is bruising at troughs during concentrate feeding; but the specific cause has now been established as a bacterial infection that involves *staphylococcus aureus*.

Symptoms
Round abscesses form and burst to produce crusty, suppurative sores on the head, especially round the eyes, but also on the face and poll.

Severe outbreaks cause dullness, loss in condition, and occasionally a nasty infection round the eye that damages the eye and often closes it completely.

Treatment
Injections of long-acting broad-spectrum antibiotics can produce spectacular cures.

Prevention
Isolate the affected sheep for treatment, and provide more trough space. Some shepherds solve the problem by floor feeding (*photo 7*), although the condition can appear independently of trough feeding.

Alopecia (wool slip)

This occasionally occurs in the odd ewe following winter shearing in housed sheep, especially if the ewe is in poor condition. The

loss of the protective wool may lead subsequently to secondary chills and pneumonia when the ewe is turned out.

Symptoms
Bald patches appear on the sides, hindquarters, back and neck, and the surrounding wool is easily plucked (*photo 8*).

Treatment
If the ewe happens to be in poor condition, get the veterinary surgeon to check the teeth in case the food intake is inadequate.

Prevention
Where the problem is persistent or when several ewes are affected, increase the quality of the ration.

67 Photosensitization

Facial eczema, yellowses, head grit

Cause
Exposure of sheep to sunlight after they have eaten certain plants or been dosed with one of several drugs.

In New Zealand the toxin of a fungus called *Phithomyces chartarum* is the most common trigger factor.

In Britain it occurs mostly in hill lambs during June and July, though it is often seen in sheep being fed on rape.

Symptoms
The head is the vulnerable area simply because the wool protection there is negligible. There occurs marked swelling of the white parts of the head and ears, often causing the ears to dangle down with the extra weight. The tips of the ears may become necrotic (dead) and drop off. The eyelids, face and lips also become swollen, and later, crusts and scabs appear (above).

Many sheep increase the damage by rubbing the head against fence posts and so on, and the ears may become crumpled.

Other exposed white areas of the body may show similar signs.

Treatment and control
Get the affected animals inside in a darkened shed or box and spray the lesions with an aerosol of broad-spectrum antibiotic to prevent secondary sepsis.

The recovery rate is high provided secondary infection does not damage the liver. If it does, jaundice and death can result.

Allergic dermatitis

Cause
The bites from midges of the *Collicoides* species which attack sheep during the warmer weather from April to October.

Symptoms
These appear only on a small percentage of the sheep that are allergic to the bites. An irritant dermatitis develops on the head, teats, udder and ventral abdomen. The condition is similar to sweet itch in horses and requires veterinary attention.

Headfly disease

Cause
A non-biting fly which swarms round sheep's heads, interrupting feeding from June to October. Horned sheep seem most susceptible, but it can occur in all types of sheep, particularly those grazing close to woodland areas.

It is very common in the Scottish borders, west Scotland and northern England.

Symptoms
The flies cause an irritation which makes the animals knock their heads on the ground or trough producing wounds which become infected and/or fly-blown.

Treatment
Local application of tetracyclene aerosols.

Prevention
Keep the flocks at least 15m from woodlands. Protect the vulnerable areas with insecticides in oil or organophosphorus cream. The old-fashioned Stockholm tar is an excellent fly repellent.

Valuable sheep can be protected by fitting a canvas hood (Marshall's Anticap) over the head. Highly effective pour-on repellents are now available.

Nasal fly disease

This is a summer disease that apparently affects sheep in many parts of England, especially in the south Midlands; so far, however, it has not been recorded in Scotland (*photo 1*).

Cause
It is caused by the larvae of a fly called *Oestrus ovis* (*see diagram*). The adult fly lays its larvae in the nostrils of the sheep.

The larvae feed on the nasal mucous membrane for up to 9 months, then migrate up into the sinuses of the head, undergoing two changes and eventually passing down again into the nostrils to be sneezed out on to the pastures, where they become adult flies.

Symptoms
The majority of infestations are by only a few larvae and show little or no sign.

With heavy infestations the affected animal shakes its head and rubs its nose on the ground.

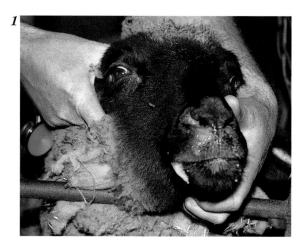

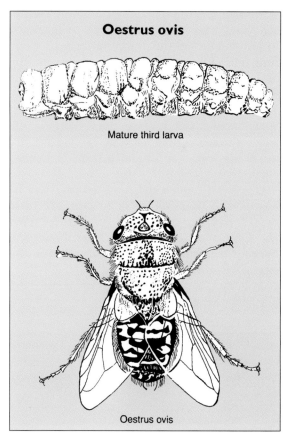

Oestrus ovis

Mature third larva

Oestrus ovis

The nostrils develop a pusy discharge, and the patient grazes less and loses weight, sneezing and coughing occasionally.

Treatment
Anthelmintics marketed under the names Ivermectin or Ivomec and Closantel are highly effective against the parasite. Note: Drugs are continually being improved so it is always wise to consult your veterinary surgeon.

Prevention
Persuade all neighbouring flocks to use the same effective anthelmintics.

68 Pneumonia

The term 'pneumonia' simply means lack of air, and the lack of air is caused by inflammation of the lungs (*photo 1*). Such lung inflammation can affect sheep at any age, although the incidence and causes of sheep respiratory disease are still vague. Without doubt the simple fact that most sheep spend the greater part of their life outdoors means that respiratory diseases are much less common in sheep than in housed animals.

Obviously, with all lung conditions in sheep, the best policy is to consult your veterinary surgeon. He, probably with the help of a scientist, will establish the correct diagnosis and will draft out the best procedure to follow.

Pneumonia can be (1) bacterial, (2) parasitic, or (3) mechanical.

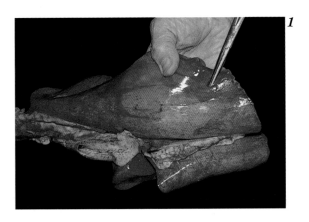
1

Bacterial pneumonia (enzootic pneumonia, pasteurellosis)

This is by far the most serious sheep lung condition. In fact pasteurellosis is one of the most important causes of economic loss to the British sheep industry.

Cause
The main germ implicated is the *Pasteurella haemolytica*, of which there are at least twelve known serotypes: one type causes acute septicaemia.

Other germs involved in sheep pneumonias are *Corynebacterium pyogenes*, *Actinobacillus lignieresi* and *Staphylococcus aureus*. Chronic pneumonias can be due to *Mycoplasma ovipneumoniae* or to lung abscesses secondary to other general infections such as tick pyaemia in lambs. But in all suspect cases it is advisable to leave it to your veterinary surgeon to establish the definitive diagnosis and to prescribe the correct course of action (*photo 2*).

Predisposing causes
The stress of driving to market or travelling in lorries very often causes it to flare up.

Keeping sheep indoors in badly ventilated buildings can also trigger it off (*photo 3*).

Sudden changes in environment, particularly in the autumn, may produce a severe outbreak, especially in lambs.

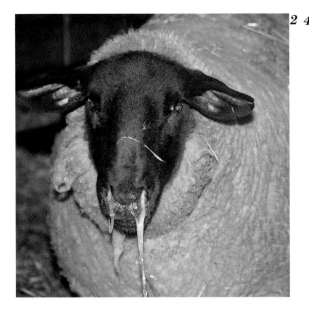

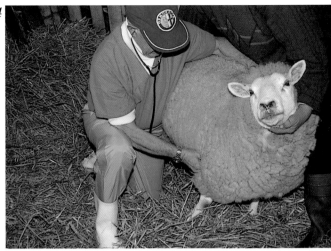

Symptoms

Enzootic pneumonia is an acute disease and the only symptom may be sudden death. Usually, however, the affected sheep run a high temperature (106° or 107°F, 41° or 41.6°C); they breathe rapidly and have a nasty, dry, hard cough. Using a stethoscope, the veterinary surgeon can detect areas of lung consolidation (*photo 4*). Infected sheep may froth at the mouth and nostrils.

Generally speaking the condition occurs in adults, mainly in the late spring, and during that time it causes septicaemia in young lambs 4–6 weeks old.

Throughout the summer and autumn it is the fattening and store lambs that seem to be the most vulnerable.

Treatment

Isolate all suspect cases immediately.

Broad-spectrum antibiotics such as tetracyclene given early have some beneficial effect in control, but the most important step is to remove any predisposing cause if at all possible. For example, if sheep are housed, improve the ventilation by providing high ridge outlets for the foul air, and remove all ground draughts, at the same time, reducing the stocking rates.

When pasteurellosis flares up in the outdoor flock, sometimes a move to a poorer and more open grazing will apparently stop the outbreak.

Prevention

Specific vaccines are available and a recent improved vaccine included in the multi-clostridial vaccines is also effective, providing an immunity to the ewes for 6 months and to the lambs for 4–6 weeks. To protect the older lambs it is necessary to vaccinate them at 3 and 4 months.

If at all possible, avoid all stress factors such as markets, housing, and sudden

Treatment

Consult your veterinary surgeon at once. He will diagnose the condition by examining samples of dung (*photo 5*) and will prescribe specific treatment.

Fortunately husk in sheep rarely becomes the serious problem that it does in calves, but nonetheless any outbreak of coughing should be investigated by your veterinary surgeon. *Routine worm dosing with modern anthelmintics does much to control lung worms.*

Mechanical pneumonia

6

All too frequently pneumonia is due to the misuse of the drenching gun (*photo 6*). **In careless or ignorant hands the dosing gun is a lethal weapon.** What happens is that the liquid worm or fluke remedies are pushed into the windpipe causing choking, drowning or, if the victim survives long enough, pneumonia.

Carbolic dips used before sales or shows are another cause.

Treatment

Casualty slaughter for food is undoubtedly the best treatment; however, carbolic dip carcases will most likely be condemned because of the smell.

Prevention

If a drenching gun has to be used, the sheep's head should be kept horizontal and the nozzle introduced gently – not into the back of the throat, but merely over the bulge at the rear of the tongue (*photo 7*).

changes in environment, particularly in the autumn when the disease is most likely to break out.

Parasitic pneumonia (husk)

Cause
Two species of lung worm – *Dictyocaulus filaria* and *Muellerius capillaris*.

Symptoms
Heavy infestations cause bronchitis. This gives rise to severe coughing which can allow parasitic or bacterial pneumonia to develop. There is usually a discharge from the eyes and nose.

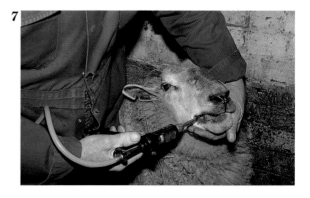

7

Fortunately, most anthelmintics can now be given hypodermically, and in my opinion, this is by far the easiest and best method. Certainly it will completely prevent the many mechanical pneumonias resulting from drenching.

Do not use carbolic dip.

69 Chronic Pneumonias

There are two chronic pneumonias of sheep, seen mainly in Iceland.

Maedi-visna

A slowly developing virus disease of both sheep and goats. The name is of Icelandic origin (*maedi* meaning 'laboured breathing' and *visna* meaning 'wasting') since the disease first developed in the housed sheep of Iceland. In maedi, the virus attacks the lungs and in visna it affects the brain; maedi is by far the most common form of the disease.

Though not yet an acute clinical problem in Britain, it was first confirmed in 1979 and breed societies conscious of its dangers have evolved a maedi-visna accredited flock scheme which is progressively gaining in popularity. The presence of the virus in a flock can be detected by blood-testing.

Symptoms

Affected sheep, usually over 3 years of age, may lag behind the flock and show fast shallow breathing and coughing after exercise. This is due to a slowly developing pneumonia.

As the disease progresses, breathing becomes more difficult and death supervenes within 3 to 7 or 8 months.

Visna, rarely seen outside Iceland, causes progressive paralysis of the hind-quarters in sheep over 2 years old.

Obviously the virus is most likely to spread by direct contact in housed sheep, though the

1

ewe's milk transmits the virus to young lambs where it remains as a permanent menace.

Treatment

There is none. Once the virus gets going affected sheep do not recover.

Prevention

If the virus is detected in a flock then all the progeny from that flock should be sold only for slaughter, thus avoiding spread to other flocks.

No satisfactory vaccine is as yet available.

Obviously in pedigree flocks the maedi-visna accredited flock scheme is the only completely satisfactory answer to the danger. It is controlled by blood testing (above).

Jaagsiekte
(Pulmonary adenomatosis)

A chronic progressive disease of the lungs. The lesion is essentially a neoplasm or growth which is contagious and transmissible. Although seen mainly in housed sheep in Iceland, it has caused losses in some British flocks, particularly those of Scottish origin, and it occurs worldwide except in Australia and New Zealand.

Cause

Essentially a neoplasm or growth thought to be due to two slow viruses working together, a herpes virus and a retro virus. The incubation period is up to 3 years. It is contagious.

Symptoms

Usually seen in adult sheep 3 to 4 years old; the disease develops slowly.

The first sign is a shallow cough and rapid breathing after being driven.

As the disease progresses breathing becomes more rapid and difficult. The animal becomes emaciated and when it attempts to graze there is a profuse watery discharge from the nostrils. If not culled the patient dies within 3 months of the early cough. The nasal discharge is infectious.

Treatment

No specific treatment is available.

Prevention

Since the incubation period of the viruses is so long, practical control is impossible. It is advisable, however, to slaughter affected animals as soon as the condition is diagnosed in order to prevent spread.

2

Pneumonia of housed sheep

A chronic type of pneumonia more or less identical to that seen in intensively housed pigs (*photo 2*).

Cause

It is caused by an organism called *Mycoplasma ovipneumoniae* (see p. 164).

Symptoms

The outstanding symptoms are chronic and persistent coughing allied to unthriftiness. Affected lambs may take up to 2 months longer to reach slaughter weight, thereby eroding the profit margin by eating excess food.

Also the lowered resistance in the lungs occasionally allows a flare-up of the acute *Pasteurella* infection (see p. 164).

Prevention

A. House the sheep for only a limited period.
B. If this is not possible, improve the ventilation.
C. Reduce the numbers in any one building.
D. There is no vaccine against the *Mycoplasma*, though it may be economical to vaccinate against the *Pasteurella* to avoid loss by death.

70 Scrapie

Scrapie is a chronic nervous disease of adult sheep (*photo 1*). It can also affect goats.

Cause
The cause is unknown, but it may be an exceptionally powerful filtrable virus which settles in the central nervous system (the brain and spinal cord) and in the spleen; it also withstands boiling.

The disease appears to be spread via direct contact, the pasture, infected premises or by congenital infection through either the ewe or the tup, though both may appear healthy and normal at the time.

It attacks when the ewes or tups are anything from 1½ to 9 years old, and it takes from 18 months to 3 years for the symptoms to appear. The peak incidence is at 3½ to 4 years.

The disease occurs in Scotland, numerous parts of England, Western Europe and the USA.

Although it can occur in all breeds, it would appear that certain breeds are more susceptible than others – scrapie is most commonly seen in the Border Leicester, the Suffolk, the Cheviot, and in the half-bred crosses from these breeds. It is rare in the Scottish Blackface and in the Blackface crosses.

Some years up to 20 per cent of the flock may be affected; in other years there may be only an occasional case.

Symptoms
Clinical disease develops progressively over months or years, and several sets of symptoms have been described. Namely:

1. 'Scratching'. This is the most common syndrome. The affected animal scrapes and rubs against posts or fences (*photo 2*). It starts to nibble or bite its skin due to a marked pruritis (itching). If horned it may remove large quantities of wool. The 'nibbling

reflex' can be produced by scratching the back, rump or flanks. (A more severe form of itching may be caused by the Aujeszky's virus of pigs.)

2. Following on from the 'scratching', or independent of it, a suspect sheep may become 'hyperexcitable'. Shivering of the head and neck may cause slight 'nodding'. When chased or rounded up these nervous symptoms become more marked. Some of the sheep thus affected may become drowsy or stupid.

3. A third and less common type is where the ewe gradually loses control of the hind-quarters. It develops a pronounced swaying and eventually flops down for good (*photo 3*).

The itching interferes with eating and sleeping and leads to a marked loss in condition. The wool becomes lighter in colour. Affected sheep all die within 1 to 6 months of the onset of the symptoms.

Treatment
There is no treatment, and affected animals should be removed for slaughter as soon as the symptoms appear.

Prevention and control
This is limited to the removal of affected sheep from the flock. Ideally the progeny of affected cases should also be culled, as well as close relatives. Obviously this latter course is seldom economically practical. Any suspect cases should be lambed in isolation and the foetal membranes destroyed.

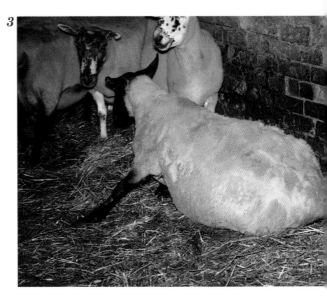
3

Scrapie and BSE
There has been a great deal of discussion as to the relationship between scrapie in sheep and bovine spongiform encephalopathy in cattle (BSE). Recent outbreaks of BSE in cattle are thought to be due to the inclusion of sheep offal in cattle food concentrates. The evidence is strong enough for a ban to have been imposed on the use of ovine products in this way. If it can be proved conclusively that sheep offal is responsible, it would substantiate the theory that scrapie is caused by a filtrable virus. In the meantime, all cases of suspected scrapie have to be reported to the Ministry of Agriculture.

71 Uncommon Sheep Conditions

Tuberculosis

A rare disease in sheep, producing a gradual loss of weight. Coughing is not a feature. Post-mortem shows widespread TB nodules in the lymph glands and peritoneum.

Ringworm

Sometimes seen in sheep housed in buildings that were previously used for infected calves – in *photo 1* the ringworms are on the left side of the ewe.

Treatment
Local treatment only – scrub off the scabs and apply an antifungal dressing. Note: Griseofulvin must *not* be given to pregnant ewes. In fact it is of doubtful value to any sheep.

Ked infestation

Used to be common in late autumn and winter particularly in hill sheep. It was rare for a time owing to a drop in the population of the sheep ked (*Melophagus ovinus*) – a dark brown wingless fly. The keds had been reduced by regular dipping in October/ November combined with the measures used for tick control. Lately, however, the parasites have increased in numbers.

Symptoms
Since the ked spends its entire life cycle on the sheep it makes the sheep restless. The ewe rubs itself on fences and damages the fleece and skin. Affected lambs only gain weight slowly.

Treatment and control
Regular dipping.

Pizzle rot (posthitis)

This is not by any means common but it is sometimes seen in flocks fed on excess green vegetables.

The ram's sheath and the tip of the penis become ulcerated and necrotic, and the ewes he serves may develop a vaginitis.

Cause
Probably a virus with secondary bacterial infection.

Treatment

Ideally cull the affected animal, though if the ram is valuable he can be isolated and treated with local antibiotics and long-acting injections of tetracycline.

Accidents

The most common I have encountered are wounds from sheep-worrying by dogs, damage from rams fighting, and broken legs in lambs due to rough lambing.

Sheep-worrying

An ever-present hazard on farms close to towns or cities.

Treatment

Always call your veterinary surgeon. Not only will he prescribe the best possible treatment, but his evidence will be vital in any insurance claim.

Damage from rams fighting

Treatment

Again, one for your veterinary surgeon. The case illustrated in *photo 2* required an eye removal under general anaesthesia.

Broken legs in lambs

Due to rough lambing.

Treatment

An immediate visit to the veterinary surgeon. The young animal's bones heal rapidly under a 10–14 day plaster (*photo 3*).

2 *3*

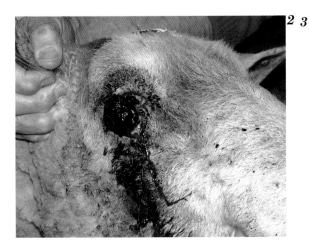
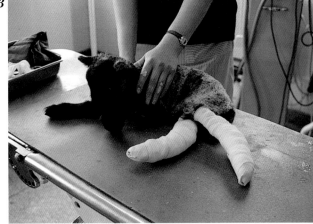

Health and Management

72 Protecting the Indoor Ewe and Lamb

Living outdoors has a great deal to commend it, and for countless years sheep, by doing so, have avoided the disease problems of intensification. In recent years, however, there has been a move towards housing ewes and lambs (*photo 1*) and naturally this has produced special complications. It is vital, therefore, to be aware of the dangers and to take every possible step to control and prevent them.

Worms present no problem in lambs housed from birth (this is obviously one great advantage of the system), but they can cause deaths in older ewes, especially when they are not being very well fed.

The routine prevention in all housed enterprises is to dose *all* the ewes when they are first housed, and to get your veterinary surgeon to check dung samples a month after dosing.

Coccidiosis may start in the housed lamb about a month after birth, causing a rapid loss in condition.

Routine prevention involves dung checks on the ewes, and possibly preventative medication of the feed and drinking water.

Lice are the chief cause of itching among housed ewes, especially if the ewes are on a

low plane of nutrition.

To avoid this trouble, dip all ewes before housing, and if the food is sparse, dip them again later on, in January or in February.

Pregnancy toxaemia is prevented in the same way as for outdoor sheep, that is, by a steady increase in feeding and condition during the 8 weeks before lambing; but control is made easier in housed ewes by grouping and penning the ewes according to their lambing date. If the lambing dates are not known, then group them by weight. All ewes that appear to be losing condition should be fed and housed separately.

Pneumonia and bronchitis. Provided the diet is adequate and consistent, the most important successful preventative of pneumonia and bronchitis is correct ventilation in the house; so before you consider housing your ewes, get the house 'vetted' in the true sense of the word. Have no fear; your veterinary surgeon will not hesitate to consult a specialist if he has any doubts.

Copper poisoning is a possible hazard when concentrates are fed to housed sheep. Keep veterinary supervision of the copper content of your ration and *never, under any circumstances, feed minerals containing copper*. (See Copper Poisoning, p. 149.)

Acidosis is a condition where the housed ewe goes off her food almost completely. It is seen chiefly when barley is first fed as a concentrate and insufficient fibre is available. Obviously, the simple answer here is to get the ewes inside and used to eating hay before starting the 8-week build-up of concentrates containing as little barley as possible or none at all. It has been my experience that barley is a menace to the pregnant and lactating ewe, consistently producing metabolic and digestive disturbances. (See Acidosis, pp. 26–7.)

Inflammation of the rumen or first stomach is seen in the growing lamb fed on barley concentrates with little or no roughage. It causes colicky pains and death. The answer? Don't feed barley and, *as a routine, always provide ad lib good quality hay*. With ad lib hay you'll probably get away with feeding a percentage of barley, but you are much better without it.

Infectious diseases generally. When planning the house, aim to have the ewes penned in small numbers. The ideal is around 12 ewes per pen, with 25 the maximum.

Then vaccinate the ewes against all the enterotoxaemic diseases, pasteurellosis and against tetanus, blackleg and black disease, with a booster dose of the vaccine or vaccines immediately before lambing.

Orf. Although, as stated in the 'orf' chapter, the vaccine is not 100 per cent effective, it does avoid an outbreak of the malignant type which can take three-quarters of the lamb crop. So I advise routine orf vaccination in all housed ewes. The vaccination is cheap, and your veterinary surgeon will supply it and instruct in its application. (See Orf, p. 158.)

Eczema around the eyes is due to a contagious bacterium, probably a *Staphylococcus* that seems to spread when the ewes are troughing. Control it by penning in small numbers, and prompt isolation and treatment with broad-spectrum antibiotic injections (see p. 161).

Abortion is the greatest potential hazard to the pregnant housed ewe. Apart from the traumatic miscarriage, there are two main types:

Enzootic abortion is now found all over Britain. Consequently, vaccination against it is an absolute must in all housed ewes. Even vaccinated ewes may occasionally abort. If they do, they should be penned separately at subsequent lambings.

Vibrionic abortion is due to *Campylobacter (Vibrio) fetus*, an organism which is picked up by the mouth, and is not spread by the ram. So far, there is no satisfactory vaccine against this, but there is every prospect that one may soon be available. In the meantime, the penning of the ewes is the most effective control weapon.

One important point: recovered ewes have a strong immunity and should be kept – not discarded. Also, as with enzootic abortion, they should be penned separately at subsequent lambings (*see* pages 83–6 for other causes).

Salmonellosis. There are several types of the germ which can cause abortion, enteritis or death in the adults and septicaemia in the lambs. So far, the sheep infection is apparently limited to certain areas, and the simple preventative measure in those areas is *not to house sheep under any circumstances*.

Contagious ophthalmia (i.e. the eye infection caused by *Rickettsia*). Obviously, this is much more likely to spread when the ewes and lambs are bunched together. To control, isolate at once and give immediate treatment with antibiotic (preferably chloramphenicol) preparations.

Mastitis is much more likely to occur in housed ewes, and it is more difficult to spot the early symptoms. It is seen chiefly when the lambs are weaned at about 6 weeks old. One way of preventing mastitis is to wean gradually, allowing the lambs to suck for 10 minutes once a day for a week or two, then once every second day. As in cattle, there is no efficient vaccine.

Foot rot. If the ewe's feet are carefully treated before housing, this disease is not likely to become a major problem. Nonetheless I advise the use of foot rot vaccine, as well as feet care, before the sheep are penned up.

For the new-born and young lambs the two greatest hazards of housing are navel infections and *E. coli*, though to some extent the losses from these in housed lambs are balanced by the number saved by being born in comparative warmth.

Navel infections lead to septicaemia and joint-ill. The most effective control measures are, first of all, to try to lamb each pen down in as short a period as possible. This avoids the build-up of infection on the pen floors. Secondly, dress the navels of the lambs with iodine or antibiotics as soon after birth as possible.

E. coli infection. Sometimes this causes diarrhoea when the lambs are 2 or 3 days old, but usually what happens is that the lamb stops sucking about 12 to 36 hours after birth and dies a few hours later. The control? Prompt treatment by oral and parental antibiotics. This calls for close liaison with your veterinary surgeon, who will have your *coli* carefully typed. Also, as with navel infection prevention, aim to lamb the entire pen in as short a period as possible to prevent disease build-up.

Where there is a history of the disease, dose each lamb by the mouth with an appropriate antibiotic before the lamb is 12 hours old.

In all cases – as soon as evidence of navel infection or *E. coli* appears – remove the ewes still to lamb to a clean pen immediately.

It must be quite obvious that lambing ewes indoors is *not* an enterprise to be undertaken lightly. The owner, the shepherd and the veterinary surgeon must play their parts, but the vet *must* be in on the entire project right from the start. If, as so often happens, he is consulted only after trouble appears then losses can be catastrophic.

In the housing of all sheep, particularly the breeding ewe, preventative medicine planned and prescribed by the veterinary surgeon is the most essential economic factor in the entire enterprise.

Winter housing

In certain parts of the country there has been a move towards housing the ewe flock over a 3- to 4-month winter period. This allows farmers to take two crops of wool and at the same time cuts out lamb losses due to inclement weather. With such enterprises, the disease precautions outlined above in this chapter are even more important. In fact they are absolutely vital.

Advantages of winter housing
- Provides the chance to shear ewes before lambing which should raise the lambing average and also the lamb weight at birth.
- Gets the ewes into better condition before, during and after lambing which should increase the lambs-reared average.
- Greatly improves the lot of the shepherd and allows much closer supervision.
- Stops the grazing being continually 'poached', especially in bad weather.
- Makes control of feeding easier.
- Insures against winter rustling and dog worrying.

Disadvantages of winter housing
- Greatly increases the risk of disease

development and spread. Most farm animal diseases are due to housing in close confinement; up to the present, outdoor sheep have avoided many of them. In my opinion this disadvantage alone potentially cancels out some of the advantages.
- Considerable capital required to provide building, troughs, electricity and bedding, water, etc.
- Much more food required, especially hay.

The happy medium

By trial and error the happy medium has apparently been arrived at by confining the lambing ewes for a ***maximum of 6 to 8 weeks*** (see Hypothermia, page 57). During this limited time the disease risk, by and large, seems to disappear and the advantages far outweigh the disadvantages.

73 Hill Sheep Farming

In Scotland

The Scottish sheep industry is based on a three-tier system, and some knowledge of this system is essential in order to understand the disease control that is practised.

On the higher hills which form the top tier (*photo 1*), Blackface and Cheviot flocks are bred pure and the only 'bought in' sheep are replacement rams. The ewes remain on the hill throughout the whole of their productive life, that is, until they are 5½ years old or until they have produced their first 4 lambs. They are then sold or 'cast'.

Lambing occurs on the hill around mid-

1

April, and the lambs remain with the ewes until they are weaned in September.

The 'cast' Blackface and Cheviot ewes are sold in the autumn and are bought by sheep farmers on the middle tier, that is, farms comprising the lower slopes of a hill farm or a better hill farm at a lower altitude. There the ewes are crossed with a Border Leicester ram; this produces a greyface lamb with the blackface, and a halfbred lamb with the Cheviot.

Lambing on these marginal farms begins in the middle of March.

The lowest or third tier comprises the lowland or arable farms. There the greyface and the halfbred ewes are crossed with Oxford Down or Suffolk rams to produce early fat lambs for slaughter. On these lowland farms lambing occurs in January or February.

On the top tier the replacement ewes come from the previous year's lamb crop. The chosen ones spend 6 months – from the end of September to the beginning of April – on a lowland farm. On 1 April they are dipped in an anti-tick dip, then turned out on the part of the hill on which they were born.

Preventative medicine plan

The diagram below, prepared by Mr Alex Wilson, MRCVS, sets out the preventative routine for all hill sheep farms to follow.

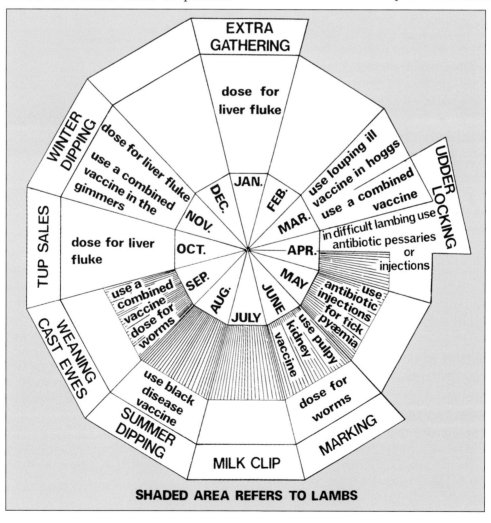

74 Recent Advances and Their Practical Commercial Potential

Commercial embryo transfer

This can be done surgically and with a high degree of success. I personally have performed several such transfers.

In my opinion the technique has limited potential in sheep and will in the future be restricted to the most valuable of pedigree animals. Obviously it should only be done by skilled veterinary surgeons or research scientists.

Artificial insemination

This has great potential and is on the increase in progressive breeding, despite the intricate anatomy of the ewes' cervix. Because of this, and since the alternative to intra-cervical entry requires the intra-abdominal use of a laparoscope, both operations comprise skilled acts of veterinary surgery. Therefore, in the interests of animal welfare, artificial insemination in sheep should be performed exclusively by the veterinary profession or by closely allied research scientists.

Controlled breeding (Ram testing, sponges, induction)

Controlled breeding has been practised by some of the best breeders for a considerable time. Progesterone sponges inserted into the vagina (*photo 1*) are the most popular method. In my opinion its potential will be restricted to these specialized flocks because of the extra labour involved.

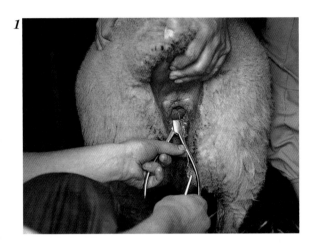

1

The vast majority of commercial flocks, however, are 'one man bands' and will remain so in the foreseeable future. For them, the use of teasers or vasectomized rams (*photo 2*) provides an economical alternative. When allowed to run with the ewes approximately a month before the start of the breeding season, the vasectomized ram or rams will undoubtedly stimulate many ewes to come into heat within 3 or 4 weeks. In other words, this is a cheap and efficient way of inducing synchronized oestrus. Also by raddling these rams it is possible to remove individual ewes for service by a specific fertile ram or for artificial insemination.

For the average sheep farmer this technique is more practical and considerably less expensive than the use of progesterone sponges.

Pregnancy diagnosis using scanners

Economically this offers great potential if only in reducing feed costs. Not only can pregnancy be confirmed, but the lamb crop can be estimated with a high degree of accuracy. In my opinion this is commercially viable for the majority of sheep farmers and will eventually be in general use by veterinary surgeons. The ultrasound scanners offer even greater potential.

Specialized feeding

In my opinion, this has a future only in top-quality pedigree flocks where capital and labour present no problems. The average commercial sheep farmer has to 'cut his cloth' according to resources, and by doing so will continue to show a reasonable financial profit. In the vast majority of business enterprises financial survival depends more on what you spend, than what you make.

My old friend the late John Cherrington put it beautifully when he told me many years ago, 'No farmer ever went bankrupt by letting old buildings fall down. They only go bust when they start building new ones.'

Condition scoring

The Meat and Livestock Commission have developed a standard scoring system for estimating the sheep's body condition. This is done by assessing the amount of fat and muscle over the spine, especially in the loin region behind and above the last rib, and also by checking the area around the eye.

In my opinion this simple scoring system has great potential for all flockmasters, especially in the purchasing of ewes and in getting the breeding flock into ideal condition for going to the ram. Unfortunately tradition will be its greatest obstacle in reaching the vast majority of sheep farmers, especially those in the more remote hill farms. Nonetheless I confidently predict that the system will be in widespread general use very soon.

A simple formula for condition scoring follows. The areas to concentrate on are the eye and the spine at the back end (i.e. the lumbar region).

The grades according to the Ministry of Agriculture are:

Grade 0
An emaciated sheep with no fat around the eye and no muscle or fat on or under the spine.

Grade 1
The spine is prominent and sharp and it is possible to get your fingers easily under the spinal bones. The eye muscles are shallow with no fat cover.

Grade 2
The spine is prominent but smooth and rounded. A little pressure is needed to get the fingers under the spine. The eye muscle can be felt but has little fat cover.

Grade 3
The ideal grade for ewes going to the tup and the standard advised for prescribing extra food. The spine has only a small elevation but feels smooth and rounded and the individual bones can be felt only with pressure. Firm pressure is needed to feel the lateral parts of the spine. The eye muscles are full and have a fair degree of fat cover.

Grade 4
Strong pressure is needed to feel the spinal processes and the lateral parts of the spine cannot be felt. The eye muscles are full and have a thick covering of fat.

Grade 5
The spine cannot be felt even with strong pressure since it is protected by layers of fat. The eye muscles are very full with thick fat cover. Usually there are large fat deposits over the rump and tail.

75 How to Keep a Clean Flock*

There is no substitute for good management. This involves efficient attention to detail and the maximum use of all available preventative measures.

The ideal is to create and rigidly keep a self-contained flock as far as possible, and to use the breeding ewes as long as their udders are sound and they continue to breed. By doing so the flock builds up a powerful resistance to all the local bacteria and parasites. In other words, a strong natural immunity is established. Also, of course, by not buying in, the danger of introducing disease is eliminated. A.I. will ensure self-contained status.

Each year the farmer should endeavour to reserve a clean pasture for the lambing ewes to graze. A clean pasture is one that has not been grazed by sheep the previous year and that has preferably carried an arable crop with the grass undersown. Ewes should also be dosed against worms at the appropriate time to prevent the spring egg-rise, as only a small number of worm eggs are passed by the adult sheep except during that period.

Internal parasites known to exist in the area should be routinely dosed against – stomach and bowel worms and, where necessary, fluke. One species of *nematodirus*, namely *N. fillicollis*, is also controlled by routine dosing. The dangerous one, *N. battus*, requires a period of freezing followed by a temperature rise to 10–12°C (50°F) for it to cause severe losses, as explained under Chapter 28, *Nematodirus* Infection. Ministry of Agriculture scientists keep a record of the soil temperatures and notify sheep farmers if special dosing is required, and this advice should be strictly followed. See under Internal Parasites.

Eliminate foot rot. Firstly, follow the recommended management routine strictly and secondly, vaccinate the flock twice a year with Clovax or Footvax.

Guard against all the metabolic diseases:

Pregnancy toxaemia
Separate all ewes with a condition score of 3 or less for special feeding. If near a scanning centre or if a scanner can be purchased or borrowed, then give even more attention to the ewes carrying twins or triplets.

Hypomagnesaemia
Use magnesium limestone instead of lime as a pasture dressing, and during the danger period feed each ewe daily with 7g (¼oz) of calcined magnesite in a succulent base such as treacle or wet beet pulp.

*All the conditions mentioned here are dealt with in greater detail elsewhere in this book.

Hypocalcaemia (lambing sickness)
See page 22.

Hypophosphaemia (phosphorus deficiency)
Use a high phosphorus general fertilizer and feed sensibly. Phosphorus deficiency, which causes rickets, double scalp and open mouth, usually occurs only when the plane of nutrition is very low. See Chapter 9.

Trace element or vitamin deficiencies
Use suitable additives in the food to prevent the following:

● *Swayback* If only odd cases occur, then 0.5 per cent copper licks may be all that is necessary to control or eliminate the problem. If numerous cases are seen, one of the modern safe copper injections or oral preparations should be used under veterinary supervision. Care must be taken to avoid feeding copper-containing minerals simultaneously, and the manufacturers' instructions should be followed to the letter. Failure to do so can produce copper poisoning, which could be much worse than the swayback.
● *Pine* If the farm is in a pine area, then each ewe should be given a cobalt bullet each year. A deficiency of the trace element cobalt prevents the formation of vitamin B_{12} in the rumen, and it is the B_{12} shortage that causes the condition.
● *Cerebrocortical necrosis* This is due to a deficiency of vitamin B_1 and can be guarded against by feeding a vitamin supplement containing thiamine (vitamin B_1).
● *White muscle disease* This can be guarded against by making sure the mineral mixture and/or vitamin supplement contains selenium and vitamin E.
● *Xerophthalmia* This is due to vitamin A deficiency, which can also predispose the animal to rickets. It can be avoided by ensuring that there is vitamin A in the supplement.

Having built up natural immunity, eliminated foot rot, controlled all the internal parasites and guarded against the metabolic disorders, the next step is to prevent all the clostridial diseases, pasteurella and erysipelas (stiff lamb disease). The clostridial multivaccines prevent enterotoxaemia (pulpy kidney), lamb dysentery, tetanus, braxy, black disease, blackleg, botulism, maligant oedema and gas gangrene. The vaccine protects the ewes for one year and the lambs for 10 weeks.

The pasteurella vaccine, which can be combined with the multi-clostridial, will protect the ewes for only 6 months and the lambs for 4–6 weeks.

When vaccinating with this multi-dose, it is a good idea to inject 2cc of Erysorb (the swine erysipelas vaccine) on the other side of the ewe's neck. This will protect the lambs against stiff lamb disease, a problem often forgotten about but one that can cause serious losses. An even greater degree of protection is provided by the new vaccine Erysorb ST (see page 56).

Before starting the breeding programme, a veterinary surgeon should test the rams. He will examine the testicles and the semen. This is important since approximately 10 per cent of rams are infertile.

Progesterone sponges are very effective for synchronizing breeding. The use of vasectomized rams ('teasers') is almost as good and much less expensive. The teaser should be allowed to run with the flock for 4–6 weeks before tupping.

It has been estimated that 70 per cent of all ewe losses, 75 per cent of all lamb losses and 95 per cent of vets' fees occur at lambing time. Bringing the ewes under cover into a well-ventilated but draught-proof environment will greatly reduce the losses. Also a second crop of wool, taken at this time, will increase the profitability. Having the lambs under cover will allow the shepherd to deal promptly with hypothermia: both the form caused by cold immediately after birth, and that due to starvation at 4–5 days old.

The most likely dangers of temporary indoor housing during lambing are coughing and contagious ophthalmia. The coughing is due to a bronchitis caused by a mycoplasma and only occasionally develops into a pneumonia. When it does, the ewe starts to breathe heavily and goes off her feed. Long-acting antibiotics, promptly given, will usually cure such cases. The dangerous type of pneumonia caused by *Pasteurella* is guarded against by the vaccine already given. Contagious ophthalmia cases should be isolated.

Inside, of course, it is vital to be as scrupulously clean as possible and practicable. The lambing pens should be swapped around each year, and plenty of clean straw bedding should be used. These simple precautions will minimize the risk of:

- *Joint-ill* (which should be further ensured against by dressing the navels at birth).
- *Mastitis* in the ewe.
- Build-up of *E. coli* and *Rotavirus* which cause scouring and watery mouth.
- *Coccidiosis*, which attacks the lambs at 4–6 weeks and causes a black diarrhoea or general unthriftiness. It is the ewe that excretes the oocysts of the Coccidia and these build up in the lambs, which have no natural resistance. It is probably wise to dose housed lambs with sulphamezathine at 4 weeks, and lambs which are turned out when a week old should be dosed at 4–6 weeks.

It is advisable to have a first-aid kit at lambing time. I advise the following:

- Several bottles of a sterile solution of calcium, phosphorus and magnesium (CPM) and a supply of 20cc syringes and needles (these should be kept suspended in a solution of the antiseptic Savlon). The CPM solution is to treat any case of hypocalcaemia (lambing sickness) which may occur before or during lambing. 20cc of the CPM solution should be injected under the skin in five different places and massaged gently till the skin is flat. A veterinary surgeon would give 20cc intravenously but I wouldn't advise any shepherd to try that.
- One or two bottles of long-acting penicillin

plus a 10cc syringe and needles (again kept suspended in Savlon solution). 6cc of the antibiotic should be injected when any internal assistance is necessary and also – very important – if the ewe retains the afterbirth. Failure to inject in the latter case will result in the ewe straining incessantly until she dies.
- Five litres (one gallon) of the non-irritant antiseptic Savlon (the veterinary surgeon will supply this).
- Half a dozen packets of soap flakes, which provide the perfect lubrication.
- Three fine (6mm) strong nylon cords.

The Savlon should be used not only to keep the syringes and needles sterile but also to add to the hot water in which the hands and arms are scrubbed before attempting any internal examination.

After washing, hands and arms should be coated with soap flakes, and both these and the water should be used copiously to cleanse the entire area around the ewe's vulva.

If the shepherd's hand is large, send for the veterinary surgeon. An alternative is to get the veterinary surgeon to train a daughter or wife using the illustrations in this book.

The important thing to remember is that one should never panic or hurry. If the ewes are apparently all right last thing at night then they will be all right for at least 6 hours.

When correcting a malpresentation always get the ewe into a suspended position with a bale of straw supporting her back and introduce the hand very slowly and carefully, between the ewe's strains. Always remember that it is easier to kill a ewe by lambing her roughly than by shooting her with a humane killer.

The other great danger at lambing time is *abortion*, and even though the incidence is not high it is vital to know what can be done to prevent it. There are four main infectious types:

Enzootic abortion
This is caused by a bacterium called *Chlamydia* and can be introduced into the

flock by purchased carrier ewes. It is spread from ewe to ewe during the 3 weeks after the abortion. Fortunately there is an ever-improving vaccine against it and when replacements are bought in it is wise to inject the ewes at tupping time; a single injection will last for 3 years. Remember the *Chlamydia* is dangerous to pregnant women, as is *Toxoplasmosis* and the vaccines against both.

Toxoplasmosis

At one time it was thought that this disease could be bought in and that it could be spread from ewe to ewe, but it is now known that the only source of infection is cat's faeces. It can therefore be prevented by keeping cats well clear of the lambing pens, the sheep's fodder and the concentrates: combined with vaccination and vermin control.

Salmonellosis

This is spread by vermin, rats mainly, so it is wise to get the rat-catcher onto the farm before bringing in the ewes.

Vibriosis

Caused by the bacterium *Campylobacter*, this is carried and spread by birds, so these should not be encouraged in the lambing houses.

It is fortunate that all these infectious abortions do not appear to impair the ewe's fertility. In fact they provide a powerful immunity so it is wise not to sell the aborted ewes.

Routine dipping controls most of the external parasitical causes of *skin diseases* – mange, lice, keds, ticks and blowfly. Wool rot and ringworm require special antifungal dips.

Ticks are a special hazard in certain areas, especially where the grazing is rough with a fair amount of rushes, heather and bracken. In order to reduce the incidence of tick-borne fever and tick pyaemia it is advisable to burn off the heather, bracken and rushes and plough up and resow the grazing. Louping ill, the other tick-borne disease, should be vaccinated against.

There are two excellent vaccines against orf – one for the ewes and one for the young lambs during the first week of life. Both are well worth the expense.

Gid can be forestalled by injecting all the farm dogs against tapeworm every 6 months. This routine should be vigorously adopted.

Periorbital eczema can be avoided by providing more trough space or by feeding from the pasture or floor.

Finally, in pedigree flocks it is an excellent idea to join the accredited Ministry of Agriculture scheme for the control of maedi-visna.

One last word: castration and/or docking should be done by the rubber ring method. The rings should not be applied, however, until the lambs are one day old with their bellies full of colostrum.

Lyme disease

One disease not so far mentioned is called Lyme disease. The reason is that it is more important as a Zoonotic infection (for its effect on humans) than for any clinical effect on sheep.

It is caused by an organism called a spirochaete which is usually transmitted by the sheep tick in the tick areas of Europe and Britain.

In humans it can produce eczematous-type skin problems, arthritis, and influenzal and nervous symptoms.

The spirochaete is called the *Borrelia burgdorferi* – a name that will interest only the veterinary student.

Anaemia in housed lambs

Lambs kept inside for up to ten days or longer may develop a degree of anaemia. This can be prevented by injecting the lambs at one day old with an intramuscular dose of iron dextrose prescribed by your veterinary surgeon. Such a precaution will ensure maximum bodyweight at weaning.

Index